Exploratory Data Analysis with Python Cookbook

Over 50 recipes to analyze, visualize, and extract insights from structured and unstructured data

Ayodele Oluleye

BIRMINGHAM—MUMBAI

Exploratory Data Analysis with Python Cookbook

Publishing Product Manager: Heramb Bhavsar
Content Development Editor: Joseph Sunil
Technical Editor: Devanshi Ayare
Copy Editor: Safis Editing
Project Coordinator: Farheen Fathima
Proofreader: Safis Editing
Indexer: Pratik Shirodkar
Production Designer: Prashant Ghare
Marketing Coordinator: Shifa Ansari

First published: June 2023

Production reference: 2300623

Published by Packt Publishing Ltd.
Livery Place
35 Livery Street
Birmingham
B3 2PB, UK.

ISBN 978-1-80323-110-5

www.packtpub.com

To my wife and daughter, I am deeply grateful for your unwavering support throughout this journey. Your love and encouragement were pillars of strength that constantly propelled me forward. Your sacrifices and belief in me have been a constant source of inspiration, and I am truly blessed to have you both by my side.

To my dad, thank you for instilling in me a solid foundation in technology right from my formative years. You exposed me to the world of technology in my early teenage years. This has been very instrumental in shaping my career in tech. To my mum (of blessed memory), thank you for your unwavering belief in my abilities and constantly nudging me to be my best self.

To PwC Nigeria, Data Scientists Network (DSN), Data Community Africa and the Young Data Professionals group (YDP), thank you for the invaluable role you played in my growth and development in the field of data science. Your unwavering support, resources, and opportunities have significantly contributed to my professional growth.

Ayodele Oluleye

Contributors

About the author

Ayodele is a data professional with nearly a decade of experience. Throughout his career, he has gained valuable experience in various domains such as strategy, data science and more recently data management. Previously, he served as a consultant at a big 4 consulting firm, where he successfully provided data-driven solutions and insights to clients. Currently, he holds a leadership position at a financial services group where he leads a dynamic data team, driving analytics initiatives to empower the organization. Beyond his professional endeavors, he is passionate about sharing his knowledge and experience. You can find him actively engaging with the data community through insightful articles on LinkedIn and speaking at industry events.

About the reviewers

Kaan Kabalak is a data scientist who especially focuses on exploratory data analysis and the implementation of machine learning algorithms in the field of data analytics. Coming from a language tutor background, he now uses his teaching skills to educate professionals of various fields. He gives lessons in data science theory, data strategy, SQL, Python programming, exploratory data analysis and machine learning. Aside from this, he helps businesses develop data strategies and build data-driven systems. He is the author of the data science blog Witful Data where he writes about various data analysis, programming and machine learning topics in a manner that is simple and understandable.

Sanjay Krishna is a seasoned data engineer with almost a decade of experience in the data domain having worked in the energy and financial sector. He has significant experience developing data models and analyses using various tools such as SQL & Python. He is also an official AWS Community Builder and is involved in developing technical content in cloud-based data systems using AWS services and providing his feedback on AWS products as a Community Builder. He is currently employed by one of the largest financial asset managers in the United States as a part of their modernization effort to move their data platform to a cloud-based solution and currently resides in Boston, Massachusetts.

Table of Contents

3

Visualizing Data in Python   47

4

Performing Univariate Analysis in Python   75

5

Performing Bivariate Analysis in Python   99

6

Performing Multivariate Analysis in Python   123

7

Analyzing Time Series Data in Python   167

8

Analysing Text Data in Python   211

9

Dealing with Outliers and Missing Values   269

10

Performing Automated Exploratory Data Analysis in Python   349...

Performing Automated Exploratory Data Analysis in Python   315

Preface

In today's data-centric world, the ability to extract meaningful insights from vast amounts of data has become a valuable skill across industries. **Exploratory Data Analysis (EDA)** lies at the heart of this process, enabling us to comprehend, visualize, and derive valuable insights from various forms of data.

This book is a comprehensive guide to EDA using the Python programming language. It provides practical steps needed to effectively explore, analyze, and visualize structured and unstructured data. It offers hands-on guidance and code for concepts, such as generating summary statistics, analyzing single and multiple variables, visualizing data, analyzing text data, handling outliers, handling missing values, and automating the EDA process. It is suited for data scientists, data analysts, researchers, or curious learners looking to gain essential knowledge and practical steps for analyzing vast amounts of data to uncover insights.

Python is an open source general-purpose programming language which is used widely for data science and data analysis, given its simplicity and versatility. It offers several libraries which can be used to clean, analyze, and visualize data. In this book, we will explore popular Python libraries (such as Pandas, Matplotlib, and Seaborn) and provide workable code for analyzing data in Python using these libraries.

By the end of this book, you will have gained comprehensive knowledge about EDA and mastered the powerful set of EDA techniques and tools required for analyzing both structured and unstructured data to derive valuable insights.

Who this book is for

Whether you are a data scientist, data analyst, researcher, or a curious learner looking to analyze structured and unstructured data, this book will appeal to you. It aims to empower you with essential knowledge and practical skills for analyzing and visualizing data to uncover insights.

It covers several EDA concepts and provides hands-on instructions on how these can be applied using various Python libraries. Familiarity with basic statistical concepts and foundational knowledge of Python programming will help you understand the content better and maximize your learning experience.

What this book covers

Chapter 1, *Generating Summary Statistics*, explores statistical concepts, such as measures of central tendency and variability, which help with effectively summarizing and analyzing data. It provides practical examples and step-by-step instructions on how to use Python libraries, such as NumPy, Pandas and

SciPy to compute measures (like the mean, median, mode, standard deviation, percentiles, and other critical summary statistics). By the end of the chapter, you will have gained the required knowledge for generating summary statistics in Python. You will also have gained the foundational knowledge required for understanding some of the more complex EDA techniques covered in other chapters.

Chapter 2, Preparing Data for EDA, focuses on the critical steps required to prepare data for analysis. Real-world data rarely come in a ready-made format, hence the reason for this very crucial step in EDA. Through practical examples, you will learn aggregation techniques such as grouping, concatenating, appending, and merging. You will also learn data-cleaning techniques, such as handling missing values, changing data formats, removing records, and replacing records. Lastly, you will learn how to transform data by sorting and categorizing it.

By the end of this chapter, you will have mastered the techniques in Python required for preparing data for EDA.

Chapter 3, Visualizing Data in Python, covers data visualization tools critical for uncovering hidden trends and patterns in data. It focuses on popular visualization libraries in Python, such as Matplotlib, Seaborn, GGPLOT and Bokeh, which are used to create compelling representations of data. It also provides the required foundation for subsequent chapters in which some of the libraries will be used. With practical examples and a step-by-step guide, you will learn how to plot charts and customize them to present data effectively. By the end of this chapter, you will be equipped with the knowledge and hands-on experience of Python's visualization capabilities to uncover valuable insights.

Chapter 4, Performing Univariate Analysis in Python, focuses on essential techniques for analyzing and visualizing a single variable of interest to gain insights into its distribution and characteristics. Through practical examples, it delves into a wide range of visualizations such as histograms, boxplots, bar plots, summary tables, and pie charts required to understand the underlying distribution of a single variable and uncover hidden patterns in the variable. It also covers univariate analysis for both categorical and numerical variables.

By the end of this chapter, you will be equipped with the knowledge and skills required to perform comprehensive univariate analysis in Python to uncover insights.

Chapter 5, Performing Bivariate Analysis in Python, explores techniques for analyzing the relationships between two variables of interest and uncovering meaningful insights embedded in them. It delves into various techniques, such as correlation analysis, scatter plots, and box plots required to effectively understand relationships, trends, and patterns that exist between two variables. It also explores the various bivariate analysis options for different variable combinations, such as numerical-numerical, numerical-categorical, and categorical-categorical. By the end of this chapter, you will have gained the knowledge and hands-on experience required to perform in-depth bivariate analysis in Python to uncover meaningful insights.

Chapter 6, Performing Multivariate Analysis in Python, builds on previous chapters and delves into some more advanced techniques required to gain insights and identify complex patterns within multiple variables of interest. Through practical examples, it delves into concepts, such as clustering analysis, principal component analysis and factor analysis, which enable the understanding of interactions among multiple variables of interest. By the end of this chapter, you will have the skills required to apply advanced analysis techniques to uncover hidden patterns in multiple variables.

Chapter 7, Analyzing Time Series Data, offers a practical guide to analyze and visualize time series data. It introduces time series terminologies and techniques (such as trend analysis, decomposition, seasonality detection, differencing, and smoothing) and provides practical examples and code on how to implement them using various libraries in Python. It also covers how to spot patterns within time series data to uncover valuable insights. By the end of the chapter, you will be equipped with the relevant skills required to explore, analyze, and derive insights from time series data.

Chapter 8, Analyzing Text Data, covers techniques for analyzing text data, a form of unstructured data. It provides a comprehensive guide on how to effectively analyze and extract insights from text data. Through practical steps, it covers key concepts and techniques for data preprocessing such as stop-word removal, tokenization, stemming, and lemmatization. It also covers essential techniques for text analysis such as sentiment analysis, n-gram analysis, topic modelling, and part-of-speech tagging. At the end of this chapter, you will have the necessary skills required to process and analyze various forms of text data to unpack valuable insights.

Chapter 9, Dealing with Outliers and Missing Values, explores the process of effectively handling outliers and missing values within data. It highlights the importance of dealing with missing values and outliers and provides step-by-step instructions on how to handle them using visualization techniques and statistical methods in Python. It also delves into various strategies for handling missing values and outliers within different scenarios. At the end of the chapter, you will have the essential knowledge of the tools and techniques required to handle missing values and outliers in various scenarios.

Chapter 10, Performing Automated EDA, focuses on speeding up the EDA process through automation. It explores the popular automated EDA libraries in Python, such as Pandas Profiling, Dtale, SweetViz, and AutoViz. It also provides hands-on guidance on how to build custom functions to automate the EDA process yourself. With step-by-step instructions and practical examples, it will empower you to gain deep insights quickly from data and save time during the EDA process.

To get the most out of this book

Basic knowledge of Python and statistical concepts is all that is needed to get the best out of this book. System requirements are mentioned in the following table:

Software/hardware covered in the book	Operating system requirements
Python 3.6+	Windows, macOS, or Linux
512GB, 8GB RAM, i5 processor (Preferred specs)	

If you are using the digital version of this book, we advise you to type the code yourself or access the code from the book's GitHub repository (a link is available in the next section). Doing so will help you avoid any potential errors related to the copying and pasting of code.

Download the example code files

You can download the example code files for this book from GitHub at `https://github.com/PacktPublishing/Exploratory-Data-Analysis-with-Python-Cookbook`. If there's an update to the code, it will be updated in the GitHub repository.

We also have other code bundles from our rich catalog of books and videos available at `https://github.com/PacktPublishing/`. Check them out!

Download the color images

We also provide a PDF file that has color images of the screenshots and diagrams used in this book. You can download it here: `https://packt.link/npXws`.

Conventions used

There are a number of text conventions used throughout this book.

`Code in text`: Indicates code words in text, database table names, folder names, filenames, file extensions, pathnames, dummy URLs, user input, and Twitter handles. Here is an example: "Create a histogram using the `histplot` method in `seaborn` and specify the data using the `data` parameter of the method."

A block of code is set as follows:

```
import numpy as np
import pandas as pd
import seaborn as sns
```

When we wish to draw your attention to a particular part of a code block, the relevant lines or items are set in bold:

```
data.shape
(30,2)
```

Any command-line input or output is written as follows:

```
$ pip install nltk
```

> **Tips or important notes**
> Appear like this.

Get in touch

Feedback from our readers is always welcome.

General feedback: If you have questions about any aspect of this book, email us at `customercare@packtpub.com` and mention the book title in the subject of your message.

Errata: Although we have taken every care to ensure the accuracy of our content, mistakes do happen. If you have found a mistake in this book, we would be grateful if you would report this to us. Please visit www.packtpub.com/support/errata and fill in the form.

Piracy: If you come across any illegal copies of our works in any form on the internet, we would be grateful if you would provide us with the location address or website name. Please contact us at `copyright@packtpub.com` with a link to the material.

If you are interested in becoming an author: If there is a topic that you have expertise in and you are interested in either writing or contributing to a book, please visit `authors.packtpub.com`.

Share Your Thoughts

Once you've read *Exploratory Data Analysis with Python Cookbook*, we'd love to hear your thoughts! Scan the QR code below to go straight to the Amazon review page for this book and share your feedback.

https://packt.link/r/1-803-23110-6

Your review is important to us and the tech community and will help us make sure we're delivering excellent quality content.

Download a free PDF copy of this book

Thanks for purchasing this book!

Do you like to read on the go but are unable to carry your print books everywhere? Is your eBook purchase not compatible with the device of your choice?

Don't worry, now with every Packt book you get a DRM-free PDF version of that book at no cost.

Read anywhere, any place, on any device. Search, copy, and paste code from your favorite technical books directly into your application.

The perks don't stop there, you can get exclusive access to discounts, newsletters, and great free content in your inbox daily

Follow these simple steps to get the benefits:

1. Scan the QR code or visit the link below

https://packt.link/free-ebook/9781803231105

2. Submit your proof of purchase

3. That's it! We'll send your free PDF and other benefits to your email directly

1

Generating Summary Statistics

In the world of data analysis, working with tabular data is a common practice. When analyzing tabular data, we sometimes need to get some quick insights into the patterns and distribution of the data. These quick insights typically provide the foundation for additional exploration and analyses. We refer to these quick insights as summary statistics. Summary statistics are very useful in exploratory data analysis projects because they help us perform some quick inspection of the data we are analyzing.

In this chapter we're going to cover some common summary statistics used on exploratory data analysis projects. We will cover the following topics in this chapter:

- Analyzing the mean of a dataset
- Checking the median of a dataset
- Identifying the mode of a dataset
- Checking the variance of a dataset
- Identifying the standard deviation of a dataset
- Generating the range of a dataset
- Identifying the percentiles of a dataset
- Checking the quartiles of a dataset
- Analyzing the **interquartile range (IQR)** of a dataset

Technical requirements

We will be leveraging the `pandas`, `numpy`, and `scipy` libraries in Python for this chapter. The code and notebooks for this chapter are available on GitHub at `https://github.com/PacktPublishing/Exploratory-Data-Analysis-with-Python-Cookbook`.

Analyzing the mean of a dataset

The mean is considered the average of a dataset. It is typically used on tabular data, and it provides us with a sense of where the center of the dataset lies. To calculate the mean, we need to sum up all the data points and divide the sum by the number of data points in our dataset. The mean is very sensitive to outliers. Outliers are unusually high or unusually low data points that are far from other data points in our dataset. They typically lead to anomalies in the output of our analysis. Since unusually high or low numbers will affect the sum of data points without affecting the number of data points, these outliers can heavily influence the mean of a dataset. However, the mean is still very useful for inspecting a dataset to get quick insights into the average of the dataset.

To analyze the mean of a dataset, we will use the mean method in the numpy library in Python.

Getting ready

We will work with one dataset in this chapter: the counts of COVID-19 cases by country from Our World in Data.

Create a folder for this chapter and create a new Python script or Jupyter Notebook file in that folder. Create a data subfolder and place the covid-data.csv file in that subfolder. Alternatively, you could retrieve all the files from the GitHub repository.

> **Note**
> Our World in Data provides COVID-19 public use data at https://ourworldindata.org/coronavirus-source-data. This is just a 5,818-row sample of the full dataset, containing data across six countries. The data is also available in the repository.

How to do it...

We will learn how to compute the mean using the numpy library:

1. Import the numpy and pandas libraries:

    ```
    import numpy as np
    import pandas as pd
    ```

2. Load the .csv into a dataframe using read_csv. Then subset the dataframe to include only relevant columns:

    ```
    covid_data = pd.read_csv("covid-data.csv")
    covid_data = covid_data[['iso_
    code','continent','location','date','total_cases','new_
    cases']]
    ```

3. Take a quick look at the data. Check the first few rows, the data types, and the number of columns and rows:

```
covid_data.head(5)
iso_code continent location date total_cases new_cases
0     AFG    Asia Afghanistan 24/02/2020 5      5
1     AFG    Asia Afghanistan 25/02/2020 5      0
2     AFG    Asia Afghanistan 26/02/2020 5      0
3     AFG    Asia Afghanistan 27/02/2020 5      0
4     AFG    Asia Afghanistan 28/02/2020 5      0

covid_data.dtypes
iso_code    object
continent   object
location    object
date object
total_cases    int64
new_cases int64

covid_data.shape
(5818, 6)
```

4. Get the mean of the data. Apply the mean method from the numpy library on the new_cases column to obtain the mean:

```
data_mean = np.mean(covid_data["new_cases"])
```

5. Inspect the result:

```
data_mean
8814.365761430045
```

That's all. Now we have the mean of our dataset.

How it works...

Most of the recipes in this chapter use the numpy and pandas libraries. We refer to pandas as pd and numpy as np. In step 2, we use read_csv to load the .csv file into a pandas dataframe and call it covid_data. We also subset the dataframe to include only six relevant columns. In step 3, we inspect the dataset using the head method in pandas to get the first five rows in the dataset. We use the dtypes attribute of the dataframe to show the data type of all columns. Numeric data has an int data type, while character data has an object data type. We inspect the number of rows and columns using shape, which returns a tuple that displays the number of rows as the first element and the number of columns as the second element.

In step 4, we apply the mean method to get the mean of the new_cases column and save it into a variable called data_mean. In step 5, we view the result of data_mean.

There's more...

The mean method in pandas can also be used to compute the mean of a dataset.

Checking the median of a dataset

The median is the middle value within a sorted dataset (ascending or descending order). Half of the data points in the dataset are less than the median, and the other half of the data points are greater than the median. The median is like the mean because it is also typically used on tabular data, and the "middle value" provides us with a sense of where the center of a dataset lies.

However, unlike the mean, the median is not sensitive to outliers. Outliers are unusually high or unusually low data points in our dataset, which can result in misleading analysis. The median isn't heavily influenced by outliers because it doesn't require the sum of all datapoints as the mean does; it only selects the middle value from a sorted dataset. To calculate the median, we sort the dataset and select the middle value if the number of datapoints is odd. If it is even, we select the two middle values and find the average to get the median. The median is a very useful statistic to get quick insights into the middle of a dataset. Depending on the distribution of the data, sometimes the mean and median can be the same number, while in other cases, they are different numbers. Checking for both statistics is usually a good idea.

To analyze the median of a dataset, we will use the median method from the numpy library in Python.

Getting ready

We will work with the COVID-19 cases again for this recipe.

How to do it...

We will explore how to compute the median using the numpy library:

1. Import the numpy and pandas libraries:

    ```
    import numpy as np
    import pandas as pd
    ```

2. Load the `.csv` into a dataframe using `read_csv`. Then subset the dataframe to include only relevant columns:

```
covid_data = pd.read_csv("covid-data.csv")
covid_data = covid_data[['iso_
code','continent','location','date','total_cases','new_
cases']]
```

3. Get the median of the data. Use the `median` method from the numpy library on the new_cases column to obtain the median:

```
data_median = np.median(covid_data["new_cases"])
```

4. Inspect the result:

```
data_median
261.0
```

Well done. We have the median of our dataset.

How it works...

Just like the computation of the mean, we used the numpy and pandas libraries to compute the median. We refer to pandas as pd and numpy as np. In step 2, we use `read_csv` to load the `.csv` file into a pandas dataframe and we subset the dataframe to include only six relevant columns. In step 3, we apply the `median` method to get the median of the new_cases column and save it into a variable called `data_median`. In step 4, we view the result of `data_median`.

There's more...

Just like the mean, the `median` method in pandas can also be used to compute the median of a dataset.

Identifying the mode of a dataset

The mode is the value that occurs most frequently in the dataset, or simply put, the most common value in a dataset. Unlike the mean and median, which must be applied to numeric values, the mode can be applied to both numeric and non-numeric values since the focus is on the frequency at which a value occurs. The mode provides quick insights into the most common value. It is a very useful statistic, especially when used alongside the mean and median of a dataset.

To analyze the mode of a dataset, we will use the `mode` method from the `stats` module in the `scipy` library in Python.

Getting ready

We will work with the COVID-19 cases again for this recipe.

How to do it...

We will explore how to compute the mode using the `scipy` library:

1. Import `pandas` and import the `stats` module from the `scipy` library:

    ```
    import pandas as pd
    from scipy import stats
    ```

2. Load the `.csv` into a dataframe using `read_csv`. Then subset the dataframe to include only relevant columns:

    ```
    covid_data = pd.read_csv("covid-data.csv")
    covid_data = covid_data[['iso_
    code','continent','location','date','total_cases','new_
    cases']]
    ```

3. Get the mode of the data. Use the `mode` method from the `stats` module on the `new_cases` column to obtain the mode:

    ```
    data_mode = stats.mode(covid_data["new_cases"])
    ```

4. Inspect the result subset of the output to extract the mode:

    ```
    data_mode
    ModeResult(mode=array([0], dtype=int64),
    count=array([805]))
    ```

5. Subset the output to extract the first element of the array, which contains the mode value:

    ```
    data_mode[0]
    array([0], dtype=int64)
    ```

6. Subset the array again to extract the value of the mode:

    ```
    data_mode[0][0]
    0
    ```

Now we have the mode of our dataset.

How it works...

To compute the mode, we used the `stats` module from the `scipy` library because a mode method doesn't exist in numpy. We refer to `pandas` as pd and `stats` from the `scipy` library as `stats`. In step 2, we use `read_csv` to load the `.csv` file into a `pandas` dataframe and we subset the dataframe to include only six relevant columns. In step 3, we apply the `mode` method to get the mode of the `new_cases` column and save it into a variable called `data_mode`. In step 4, we view the result of `data_mode`, which comes in an array. The first element of the array is another array containing the value of the mode itself and the data type. The second element of the array is the frequency count for the mode identified. In step 5, we subset to get the first array containing the mode and data type. In step 6, we subset the array again to obtain the specific mode value.

There's more...

In this recipe, we only considered the mode for a numeric column. However, the mode can also be computed for a non-numeric column. This means it can be applied to the `continent` column in our dataset to view the most frequently occurring continent in the data.

Checking the variance of a dataset

Just like we may want to know where the center of a dataset lies, we may also want to know how widely spread the dataset is, for example, how far apart the numbers in the dataset are from each other. The variance helps us achieve this. Unlike the mean, median, and mode, which give us a sense of where the center of the dataset lies, the variance gives us a sense of the spread of a dataset or the variability.

It is a very useful statistic, especially when used alongside a dataset's mean, median, and mode.

To analyze the variance of a dataset, we will use the `var` method from the numpy library in Python.

Getting ready

We will work with the COVID-19 cases again for this recipe.

How to do it...

We will compute the variance using the numpy library:

1. Import the numpy and pandas libraries:

    ```
    import numpy as np
    import pandas as pd
    ```

2. Load the .csv into a dataframe using read_csv. Then subset the dataframe to include only relevant columns:

```
covid_data = pd.read_csv("covid-data.csv")
covid_data = covid_data[['iso_
code','continent','location','date','total_cases','new_
cases']]
```

3. Obtain the variance of the data. Use the var method from the numpy library on the new_cases column to obtain the variance:

```
data_variance = np.var(covid_data["new_cases"])
```

4. Inspect the result:

```
data_variance
451321915.9280954
```

Great. We have computed the variance of our dataset.

How it works...

To compute the variance, we used the numpy and pandas libraries. We refer to pandas as pd and numpy as np. In step 2, we use read_csv to load the .csv file into a pandas dataframe, and we subset the dataframe to include only six relevant columns. In step 3, we apply the var method to get the variance of the new_cases column and save it into a variable called data_variance. In step 4, we view the result of data_variance.

There's more...

The var method in pandas can also be used to compute the variance of a dataset.

Identifying the standard deviation of a dataset

The standard deviation is derived from the variance and is simply the square root of the variance. The standard deviation is typically more intuitive because it is expressed in the same units as the dataset, for example, **kilometers (km)**. On the other hand, the variance is typically expressed in units larger than the dataset and can be less intuitive, for example, **kilometers squared (km²)**.

To analyze the standard deviation of a dataset, we will use the sd method from the numpy library in Python.

Getting ready

We will work with the COVID-19 cases again for this recipe.

How to do it...

We will compute the standard deviation using the numpy libary:

1. Import the numpy and pandas libraries:

    ```
    import numpy as np
    import pandas as pd
    ```

2. Load the .csv into a dataframe using read_csv. Then subset the dataframe to include only relevant columns:

    ```
    covid_data = pd.read_csv("covid-data.csv")
    covid_data = covid_data[['iso_
    code','continent','location','date','total_cases','new_
    cases']]
    ```

3. Obtain the standard deviation of the data. Use the std method from the numpy library on the new_cases column to obtain the standard deviation:

    ```
    data_sd = np.std(covid_data["new_cases"])
    ```

4. Inspect the result:

    ```
    data_sd
    21244.338444114834
    ```

Great. We have computed the variance of our dataset.

How it works...

To compute the standard deviation, we used the numpy and pandas libraries. We refer to pandas as pd and numpy as np. In step 2, we use read_csv to load the .csv file into a pandas dataframe, and we subset the dataframe to include only six relevant columns. In step 3, we apply the std method to get the median of the new_cases column and save it into a variable called data_sd. In step 4, we view the result of data_sd.

There's more...

The std method in pandas can also be used to compute the standard deviation of a dataset.

Generating the range of a dataset

The range also helps us understand the spread of a dataset or how far apart the dataset's numbers are from each other. It is the difference between the minimum and maximum values within a dataset. It is a very useful statistic, especially when used alongside the variance and standard deviation of a dataset.

To analyze the range of a dataset, we will use the max and min methods from the numpy library in Python.

Getting ready

We will work with the COVID-19 cases again for this recipe.

How to do it...

We will compute the range using the numpy library:

1. Import the numpy and pandas libraries:

    ```
    import numpy as np
    import pandas as pd
    ```

2. Load the .csv into a dataframe using read_csv. Then subset the dataframe to include only relevant columns:

    ```
    covid_data = pd.read_csv("covid-data.csv")
    covid_data = covid_data[['iso_
    code','continent','location','date','total_cases','new_
    cases']]
    ```

3. Compute the maximum and minimum values of the data. Use the max and min methods from the numpy library on the new_cases column to obtain this:

    ```
    data_max = np.max(covid_data["new_cases"])
    data_min = np.min(covid_data["new_cases"])
    ```

4. Inspect the result of the maximum and minimum values:

```
print(data_max,data_min)
287149 0
```

5. Compute the range:

```
data_range = data_max - data_min
```

6. Inspect the result of the maximum and minimum values:

```
data_range
287149
```

We just computed the range of our dataset.

How it works...

Unlike previous statistics where we used specific numpy methods, the range requires computation using the range formula, combining the max and min methods. To compute the range, we use the numpy and pandas libraries. We refer to pandas as pd and numpy as np. In step 2, we use read_csv to load the .csv file into a pandas dataframe and we subset the dataframe to include only six relevant columns. In step 3, we apply the max and min numpy methods to get the maximum and minimum values of the new_cases column. We then save the output into two variables, data_max and data_min. In step 4, we view the result of data_max and data_min. In step 5, we generate the range by subtracting data_min from data_max. In step 6, we view the range result.

There's more...

The max and min methods in pandas can also be used to compute the range of a dataset.

Identifying the percentiles of a dataset

The percentile is an interesting statistic because it can be used to measure the spread of a dataset and, at the same time, identify the center of a dataset. The percentile divides the dataset into 100 equal portions, allowing us to determine the values in a dataset above or below a certain limit. Typically, 99 percentiles will split your dataset into 100 equal portions. The value of the 50th percentile is the same value as the median.

To analyze the percentile of a dataset, we will use the percentile method from the numpy library in Python.

Getting ready

We will work with the COVID-19 cases again for this recipe.

How to do it...

We will compute the 60th percentile using the numpy library:

1. Import the numpy and pandas libraries:

```
import numpy as np
import pandas as pd
```

2. Load the .csv into a dataframe using read_csv. Then subset the dataframe to include only relevant columns:

```
covid_data = pd.read_csv("covid-data.csv")
covid_data = covid_data[['iso_
code','continent','location','date','total_cases','new_
cases']]
```

3. Obtain the 60th percentile of the data. Use the percentile method from the numpy library on the new_cases column:

```
data_percentile = np.percentile(covid_data["new_
cases"],60)
```

4. Inspect the result:

```
data_percentile
591.3999999999996
```

Good job. We have computed the 60th percentile of our dataset.

How it works...

To compute the 60th percentile, we used the numpy and pandas libraries. We refer to pandas as pd and numpy as np. In step 2, we use read_csv to load the .csv file into a pandas dataframe and we subset the dataframe to include only six relevant columns. In step 3, we compute the 60th percentile of the new_cases column by applying the percentile method and adding 60 as the second argument in the method. We save the output into a variable called data_percentile. In step 4, we view the result of data_percentile.

There's more...

The quantile method in pandas can also be used to compute the percentile of a dataset.

Checking the quartiles of a dataset

The quartile is like the percentile because it can be used to measure the spread and identify the center of a dataset. Percentiles and quartiles are called quantiles. While the percentile divides the dataset into 100 equal portions, the quartile divides the dataset into 4 equal portions. Typically, three quartiles will split your dataset into four equal portions.

To analyze the quartiles of a dataset, we will use the quantiles method from the numpy library in Python. Unlike percentile, the quartile doesn't have a specific method dedicated to it.

Getting ready

We will work with the COVID-19 cases again for this recipe.

How to do it...

We will compute the quartiles using the numpy library:

1. Import the numpy and pandas libraries:

    ```
    import numpy as np
    import pandas as pd
    ```

2. Load the .csv into a dataframe using read_csv. Then subset the dataframe to include only relevant columns:

    ```
    covid_data = pd.read_csv("covid-data.csv")
    covid_data = covid_data[['iso_
    code','continent','location','date','total_cases','new_
    cases']]
    ```

3. Obtain the quartiles of the data. Use the quantile method from the numpy library on the new_cases column:

    ```
    data_quartile = np.quantile(covid_data["new_cases"],0.75)
    ```

4. Inspect the result:

    ```
    data_quartile
    3666.0
    ```

Good job. We have computed the 3rd quartile of our dataset.

How it works...

Just like the percentile, we can compute quartiles using the numpy and pandas libraries. We refer to pandas as pd and numpy as np. In step 2, we use read_csv to load the .csv file into a pandas dataframe, and we subset the dataframe to include only six relevant columns. In step 3, we compute the **third quartile (Q3)** of the new_cases column by applying the quantile method and adding 0.75 as the second argument in the method. We save the output into a variable called data_quartile. In step 4, we view the result of data_quartile.

There's more...

The quantile method in pandas can also be used to compute the quartile of a dataset.

Analyzing the interquartile range (IQR) of a dataset

The **IQR** also measures the spread or variability of a dataset. It is simply the distance between the first and third quartiles. The IQR is a very useful statistic, especially when we need to identify where the middle 50% of values in a dataset lie. Unlike the range, which can be skewed by very high or low numbers (outliers), the IQR isn't affected by outliers since it focuses on the middle 50. It is also useful when we need to compute for outliers in a dataset.

To analyze the IQR of a dataset, we will use the IQR method from the stats module within the scipy library in Python.

Getting ready

We will work with the COVID-19 cases again for this recipe.

How to do it...

We will explore how to compute the IQR using the scipy library:

1. Import pandas and import the stats module from the scipy library:

    ```
    import pandas as pd
    from scipy import stats
    ```

2. Load the .csv into a dataframe using read_csv. Then subset the dataframe to include only relevant columns:

```
covid_data = pd.read_csv("covid-data.csv")
covid_data = covid_data[['iso_
code','continent','location','date','total_cases','new_
cases']]
```

3. Get the IQR of the data. Use the IQR method from the stats module on the new_cases column to obtain the IQR:

```
data_IQR = stats.iqr(covid_data["new_cases"])
```

4. Inspect the result:

```
data_IQR
3642.0
```

Now we have the IQR of our dataset.

How it works...

To compute the IQR we used the stats module from the scipy library because the IQR method doesn't exist in numpy. We refer to pandas as pd and stats from the scipy library as stats. In step 2, we use read_csv to load the .csv file into a pandas dataframe and we subset the dataframe to include only six relevant columns. In step 3, we apply the IQR method to get the IQR of the new_cases column and save it into a variable called data_IQR. In step 4, we view the result of data_IQR.

2

Preparing Data for EDA

Before exploring and analyzing tabular data, we sometimes will be required to prepare the data for analysis. This preparation can come in the form of data transformation, aggregation, or cleanup. In Python, the pandas library helps us to achieve this through several modules. The preparation steps for tabular data are never a one-size-fits-all approach. They are typically determined by the structure of our data, that is, the rows, columns, data types, and data values.

In this chapter, we will focus on common data preparation techniques required to prepare our data for EDA:

- Grouping data
- Appending data
- Concatenating data
- Merging data
- Sorting data
- Categorizing data
- Removing duplicate data
- Dropping data rows and columns
- Replacing data
- Changing a data format
- Dealing with missing values

Technical requirements

We will leverage the pandas library in Python for this chapter. The code and notebooks for this chapter are available on GitHub at https://github.com/PacktPublishing/Exploratory-Data-Analysis-with-Python-Cookbook.

Grouping data

When we group data, we are aggregating the data by category. This can be very useful especially when we need to get a high-level view of a detailed dataset. Typically, to group a dataset, we need to identify the column/category to group by, the column to aggregate by, and the specific aggregation to be done. The column/category to group by is usually a categorical column while the column to aggregate by is usually a numeric column. The aggregation to be done can be a count, sum, minimum, maximum, and so on. We can also perform aggregation such as count directly on the categorical column we group by

In pandas, the groupby method helps us group data.

Getting ready

We will work with one dataset in this chapter – the **Marketing Campaign** data from Kaggle.

Create a folder for this chapter and create a new Python script or Jupyter notebook file in that folder. Create a data subfolder and place the marketing_campaign.csv file in that subfolder. Alternatively, you can retrieve all the files from the GitHub repository.

> **Note**
>
> Kaggle provides the *Marketing Campaign* data for public use at https://www.kaggle.com/datasets/imakash3011/customer-personality-analysis. In this chapter, we use both the full dataset and samples of the dataset for the different recipes. The data is also available in the repository. The data in Kaggle appears in a single-column format, but the data in the repository was transformed into a multiple-column format for easy usage in pandas.

How to do it...

We will learn how to group data using the pandas library:

1. Import the pandas library:

    ```
    import pandas as pd
    ```

2. Load the .csv file into a dataframe using read_csv. Then, subset the dataframe to include only relevant columns:

    ```
    marketing_data = pd.read_csv("data/marketing_campaign.
    csv")

    marketing_data = marketing_data[['ID','Year_
    Birth', 'Education','Marital_
    Status','Income','Kidhome', 'Teenhome', 'Dt_
    ```

```
Customer',                    'Recency','NumStorePurchases',
'NumWebVisitsMonth']]
```

3. Inspect the data. Check the first few rows and use transpose (T) to show more information. Also, check the data types as well as the number of columns and rows:

```
marketing_data.head(2).T
                 0      1
ID       5524    2174
Year_Birth       1957    1954
Education        Graduation      Graduation
...      ...             ...
NumWebVisitsMonth        7     5


marketing_data.dtypes
ID      int64
Year_Birth       int64
Education        object
...              ...
NumWebVisitsMonth        int64


marketing_data.shape
(2240, 11)
```

4. Use the groupby method in pandas to get the average number of store purchases of customers based on the number of kids at home:

```
marketing_data.groupby('Kidhome')['NumStorePurchases'].
mean()
Kidhome
0       7.2173240525908735
1       3.863181312569522
2       3.4375
```

That's all. Now, we have grouped our dataset.

How it works...

All of the recipes in this chapter use the `pandas` library for data transformation and manipulation. We refer to pandas as pd in *step 1*. In *step 2*, we use `read_csv` to load the `.csv` file into a pandas dataframe and call it `marketing_data`. We also subset the dataframe to include only 11 relevant columns. In *step 3*, we inspect the dataset using the `head` method to see the first two rows in the dataset; we also use `transform (T)` along with `head` to transform the rows into columns, due to the size of the data (i.e., it has many columns). We use the `dtypes` attribute of the dataframe to show the data types of all columns. Numeric data has `int` and `float` data types while character data has the `object` data type. We inspect the number of rows and columns using `shape`, which returns a tuple that displays the number of rows as the first element and the number of columns as the second element.

In *step 4*, we apply the `groupby` method to get the average number of store purchases of customers based on the number of kids at home. Using the `groupby` method, we group by `Kidhome`, then we aggregate by `NumStorePurchases`, and finally, we use the `mean` method as the specific aggregation to be performed on `NumStorePurchases`.

There's more...

Using the `groupby` method in `pandas`, we can group by multiple columns. Typically, these columns only need to be presented in a Python list to achieve this. Also, beyond the mean, several other aggregation methods can be applied, such as max, min, and median. In addition, the `agg` method can be used for aggregation; typically, we will need to provide specific numpy functions to be used. Custom functions for aggregation can be applied through the `apply` or `transform` method in pandas.

See also

Here is an insightful article by Dataquest on the `groupby` method in `pandas`: https://www.dataquest.io/blog/grouping-data-a-step-by-step-tutorial-to-groupby-in-pandas/.

Appending data

Sometimes, we may be analyzing multiple datasets that have a similar structure or samples of the same dataset. While analyzing our datasets, we may need to append them together into a new single dataset. When we append datasets, we stitch the datasets along the rows. For example, if we have 2 datasets containing 1,000 rows and 20 columns each, the appended data will contain 2,000 rows and 20 columns. The rows typically increase while the columns remain the same. The datasets are allowed to have a different number of rows but typically should have the same number of columns to avoid errors after appending.

In `pandas`, the `concat` method helps us append data.

Getting ready

We will continue working with the *Marketing Campaign* data from Kaggle. We will work with two samples of that dataset.

Place the `marketing_campaign_append1.csv` and `marketing_campaign_append2.csv` files in the data subfolder created in the first recipe. Alternatively, you could retrieve all the files from the GitHub repository.

How to do it...

We will explore how to append data using the `pandas` library:

1. Import the `pandas` library:

    ```
    import pandas as pd
    ```

2. Load the `.csv` files into a dataframe using `read_csv`. Then, subset the dataframes to include only relevant columns:

    ```
    marketing_sample1 = pd.read_csv("data/marketing_campaign_
    append1.csv")
    marketing_sample2 = pd.read_csv("data/marketing_campaign_
    append2.csv")
    marketing_sample1 = marketing_sample1[['ID', 'Year_
    Birth','Education','Marital_Status','Income',
    'Kidhome','Teenhome','Dt_Customer',
    'Recency','NumStorePurchases', 'NumWebVisitsMonth']]

    marketing_sample2 = marketing_sample2[['ID', 'Year_
    Birth','Education','Marital_Status','Income',
    'Kidhome','Teenhome','Dt_Customer',
    'Recency','NumStorePurchases', 'NumWebVisitsMonth']]
    ```

3. Take a look at the two datasets. Check the first few rows and use `transpose` (T) to show more information:

    ```
    marketing_sample1.head(2).T
              0      1
    ID      5524   2174
    Year_Birth     1957    1954
    ...      ...     ...
    ```

```
NumWebVisitsMonth      7     5
```

```
marketing_sample2.head(2).T
        0      1
ID     9135    466
Year_Birth    1950    1944

...      ...      ...

NumWebVisitsMonth      8     2
```

4. Check the data types as well as the number of columns and rows:

```
marketing_sample1.dtypes
ID      int64
Year_Birth      int64

...      ...

NumWebVisitsMonth      int64


marketing_sample2.dtypes
ID      int64
Year_Birth      int64

...      ...

NumWebVisitsMonth      int64


marketing_sample1.shape
(500, 11)
marketing_sample2.shape
(500, 11)
```

5. Append the datasets. Use the concat method from the pandas library to append the data:

```
appended_data = pd.concat([marketing_sample1, marketing_
sample2])
```

6. Inspect the shape of the result and the first few rows:

```
appended_data.head(2).T
        0      1
ID     5524    2174
```

```
Year_Birth        1957      1954
Education       Graduation    Graduation
Marital_Status     Single     Single
Income       58138.0    46344.0
Kidhome      0     1
Teenhome     0     1
Dt_Customer      04/09/2012      08/03/2014
Recency      58    38
NumStorePurchases      4    2
NumWebVisitsMonth      7    5


appended_data.shape
(1000, 11)
```

Well done! We have appended our datasets.

How it works...

We import the pandas library and refer to it as pd in *step 1*. In *step 2*, we use read_csv to load the two .csv files to be appended into pandas dataframes. We call the dataframes marketing_sample1 and marketing_sample2 respectively. We also subset the dataframes to include only 11 relevant columns. In *step 3*, we inspect the dataset using the head method to see the first two rows in the dataset; we also use transform (T) along with head to transform the rows into columns due to the size of the data (i.e., it has many columns). In *step 4*, we use the dtypes attribute of the dataframe to show the data types of all columns. Numeric data has int and float data types while character data has the object data type. We inspect the number of rows and columns using shape, which returns a tuple that displays the number of rows and columns respectively.

In *step 5*, we apply the concat method to append the two datasets. The method takes in the list of dataframes as an argument. The list is the only argument required because the default setting of the concat method is to append data. In *step 6*, we inspect the first few rows of the output and its shape.

There's more...

Using the concat method in pandas, we can append multiple datasets beyond just two. All that is required is to include these datasets in the list, and then they will be appended. It is important to note that the datasets must have the same columns.

Concatenating data

Sometimes, we may need to stitch multiple datasets or samples of the same dataset by columns and not rows. This is where we concatenate our data. While appending stitches rows of data together, concatenating stitches columns together to provide a single dataset. For example, if we have 2 datasets containing 1,000 rows and 20 columns each, the concatenated data will contain 1,000 rows and 40 columns. The columns typically increase while the rows remain the same. The datasets are allowed to have a different number of columns but typically should have the same number of rows to avoid errors after concatenating.

In pandas, the concat method helps us concatenate data.

Getting ready

We will continue working with the *Marketing Campaign* data from Kaggle. We will work with two samples of that dataset.

Place the marketing_campaign_concat1.csv and marketing_campaign_concat2. csv files in the data subfolder created in the first recipe. Alternatively, you can retrieve all the files from the GitHub repository.

How to do it...

We will explore how to concatenate data using the pandas library:

1. Import the pandas library:

    ```
    import pandas as pd
    ```

2. Load the .csv files into a dataframe using read_csv:

    ```
    marketing_sample1 = pd.read_csv("data/marketing_campaign_
    concat1.csv")
    marketing_sample2 = pd.read_csv("data/marketing_campaign_
    concat2.csv")
    ```

3. Take a look at the two datasets. Check the first few rows and use transpose (T) to show more information:

    ```
    marketing_sample1.head(2).T
          0      1
    ID      5524    2174
    Year_Birth    1957      1954
    Education    Graduation    Graduation
    ```

```
Marital_Status      Single      Single
Income      58138.0      46344.0


marketing_sample2.head(2).T
     0     1
NumDealsPurchases       3      2
NumWebPurchases      8      1
NumCatalogPurchases      10      1
NumStorePurchases       4      2
NumWebVisitsMonth       7      5
```

4. Check the data types as well as the number of columns and rows:

```
marketing_sample1.dtypes
ID      int64
Year_Birth      int64
Education      object
Marital_Status      object
Income      float64


marketing_sample2.dtypes
NumDealsPurchases      int64
NumWebPurchases      int64
NumCatalogPurchases      int64
NumStorePurchases      int64
NumWebVisitsMonth      int64


marketing_sample1.shape
(2240, 5)
marketing_sample2.shape
(2240, 5)
```

5. Concatenate the datasets. Use the concat method from the pandas library to concatenate the data:

```
concatenated_data = pd.concat([marketing_sample1,
marketing_sample2], axis = 1)
```

6. Inspect the shape of the result and the first few rows:

```
concatenated_data.head(2).T
     0      1
ID       5524     2174
Year_Birth      1957      1954
Education      Graduation      Graduation
Marital_Status      Single      Single
Income      58138.0      46344.0
NumDealsPurchases      3      2
NumWebPurchases      8      1
NumCatalogPurchases      10      1
NumStorePurchases      4      2
NumWebVisitsMonth      7      5


concatenated_data.shape
(2240, 10)
```

Awesome! We have concatenated our datasets.

How it works...

We import the pandas library and refer to it as pd in *step 1*. In *step 2*, we use read_csv to load the two .csv files to be concatenated into pandas dataframes. We call the dataframes marketing_ sample1 and marketing_sample2 respectively. In *step 3*, we inspect the dataset using head(2) to see the first two rows in the dataset; we also use transform (T) along with head to transform the rows into columns due to the size of the data (i.e., it has many columns). In *step 4*, we use the dtypes attribute of the dataframe to show the data types of all columns. Numeric data has int and float data types while character data has the object data type. We inspect the number of rows and columns using shape, which returns a tuple that displays the number of rows and columns respectively.

In *step 5*, we apply the concat method to concatenate the two datasets. Just like when appending, the method takes in the list of dataframes as an argument. However, it takes an additional argument for the axis parameter. The value 1 indicates that the axis refers to columns. The default value is typically 0, which refers to rows and is relevant for appending datasets. In *step 6*, we check the first few rows of the output as well as the shape.

There's more...

Using the `concat` method in `pandas`, we can concatenate multiple datasets beyond just two. Just like appending, all that is required is to include these datasets in the list and the axis value, which is typically 1 for concatenation. It is important to note that the datasets must have the same number of rows.

See also

You can read this insightful article by Dataquest on concatenation: `https://www.dataquest.io/blog/pandas-concatenation-tutorial/`.

Merging data

Merging sounds a bit like concatenating our dataset; however, it is quite different. To merge datasets, we need to have a common field in both datasets on which we can perform a merge.

If you are familiar with the SQL or join commands, then you are probably familiar with merging data. Usually, data from relational databases will require merging operations. Relational databases typically contain tabular data and account for a significant proportion of data found in many organizations. Some key concepts to note when doing merge operations include the following:

- **Join key column**: This refers to the common column within both datasets in which there are matching values. This is typically used to join the datasets. The columns do not need to have the same name; they only need to have matching values within the two datasets.

- **Type of join**: There are different types of join operations that can be performed on datasets:

    - **Left join**: We retain all the rows in the left dataframe. Values in the right dataframe that do not match the values in the left dataframe are added as empty/**Not a Number** (**NaN**) values in the result. The matching is done based on the matching/join key column.

    - **Right join**: We retain all the rows in the right dataframe. Values in the left dataframe that do not match the values in the right dataframe are added as empty/NaN values in the result. The matching is done based on the matching/join key column.

    - **Inner join**: We retain only the common values in both the left and right dataframes in the result – that is, we do not return empty/NaN values.

- **Outer join/full outer join**: We retain all the rows for the left and right dataframes. If the values do not match, NaN is added to the result.

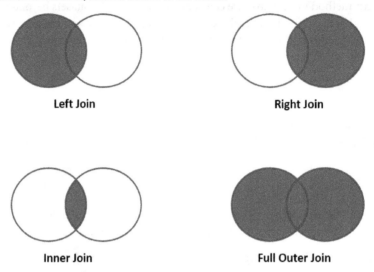

Figure 2.1 – Venn diagrams illustrating different types of joins

In pandas, the merge method helps us to merge dataframes.

Getting ready

We will continue working with the *Marketing Campaign* data from Kaggle. We will work with two samples of that dataset.

Place the marketing_campaign_merge1.csv and marketing_campaign_merge2. csv files in the data subfolder created in the first recipe. Alternatively, you can retrieve all the files from the GitHub repository.

How to do it...

We will merge datasets using the pandas library:

1. Import the pandas library:

    ```
    import pandas as pd
    ```

2. Load the .csv files into a dataframe using read_csv:

    ```
    marketing_sample1 = pd.read_csv("data/marketing_campaign_
    merge1.csv")
    ```

```
marketing_sample2 = pd.read_csv("data/marketing_campaign_
merge2.csv")
```

3. Take a look at the two datasets. Check the first few rows through the head method. Also, check the number of columns and rows:

```
marketing_sample1.head()
      ID  Year_Birth  Education
0   5524  1957        Graduation
1   2174  1954        Graduation
2   4141  1965        Graduation
3   6182  1984        Graduation
4   5324  1981           PhD


      ID    Marital_Status    Income
0   5524      Single        58138.0
1   2174      Single        46344.0
2   4141      Together      71613.0
3   6182      Together      26646.0
4   5324      Married       58293.0


marketing_sample1.shape
(2240, 3)
marketing_sample2.shape
(2240, 3)
```

4. Merge the datasets. Use the merge method from the pandas library to merge the datasets:

```
merged_data = pd.merge(marketing_sample1,marketing_
sample2,on = "ID")
```

5. Inspect the shape of the result and the first few rows:

```
merged_data.head()
      ID    Year_Birth    Education    Marital_
Status    Income
0   5524    1957      Graduation    Single    58138.0
1   2174    1954      Graduation    Single    46344.0
2   4141    1965      Graduation    Together    71613.0
```

```
3      6182      1984      Graduation      Together      26646.0
4      5324      1981      PhD        Married      58293.0
merged_data.shape
(2240, 5)
```

Great! We have merged our dataset.

How it works...

We import the pandas library and refer to it as pd in *step 1*. In *step 2*, we use read_csv to load the two .csv files to be merged into pandas dataframes. We call the dataframes marketing_sample1 and marketing_sample2 respectively. In *step 3*, we inspect the dataset using head() to see the first five rows in the dataset. We inspect the number of rows and columns using shape, which returns a tuple that displays the number of rows and columns respectively.

In *step 4*, we apply the merge method to merge the two datasets. We provide four arguments for the merge method. The first two arguments are the dataframes we want to merge, the third specifies the key or common column upon which a merge can be achieved. The merge method also has a how parameter. This parameter specifies the type of join to be used. The default parameter of this argument is an inner join.

There's more...

Sometimes, the common field in two datasets may have a different name. The merge method allows us to address this through two arguments, left_on and right_on. left_on specifies the key on the left dataframe, while right_on is the same thing on the right dataframe.

See also

You can check out this useful resource by *Real Python* on merging data in pandas: https://realpython.com/pandas-merge-join-and-concat/.

Sorting data

When we sort data, we arrange it in a specific sequence. This specific sequence typically helps us to spot patterns very quickly. To sort a dataset, we usually must specify one or more columns to sort by and specify the order to sort by (ascending or descending order).

In pandas, the sort_values method can be used to sort a dataset.

Getting ready

We will work with the Marketing Campaign data (https://www.kaggle.com/datasets/imakash3011/customer-personality-analysis) for this recipe. Alternatively, you can retrieve this from the GitHub repository.

How to do it...

We will sort data using the pandas library:

1. Import the pandas library:

    ```
    import pandas as pd
    ```

2. Load the .csv file into a dataframe using read_csv. Then, subset the dataframe to include only relevant columns:

    ```
    marketing_data = pd.read_csv("data/marketing_campaign.
    csv")
    ```

    ```
    marketing_data = marketing_data[['ID','Year_
    Birth', 'Education','Marital_
    Status','Income','Kidhome', 'Teenhome', 'Dt_
    Customer',                'Recency','NumStorePurchases',
    'NumWebVisitsMonth']]
    ```

3. Inspect the data. Check the first few rows and use transpose (T) to show more information. Also, check the data types as well as the number of columns and rows:

    ```
    marketing_data.head(2).T
                      0      1
    ID       5524     2174
    Year_Birth      1957       1954
    Education      Graduation      Graduation
    ...       ...       ...
    NumWebVisitsMonth       7       5
    ```

    ```
    marketing_data.dtypes
    ID      int64
    Year_Birth       int64
    Education      object
    ...       ...
    ```

```
NumWebVisitsMonth      int64

marketing_data.shape
(2240, 11)
```

4. Sort customers based on the number of store purchases in descending order:

```
sorted_data = marketing_data.sort_
values('NumStorePurchases', ascending=False)
```

5. Inspect the result. Subset for relevant columns:

```
sorted_data[['ID','NumStorePurchases']]
       ID    NumStorePurchases
1187    9855      13
803     9930      13
1144     819      13
286    10983      13
1150    1453      13

...     ...      ...

164     8475     0
2214    9303     0
27      5255    0
1042   10749      0
2132   11181      0
```

Great! We have sorted our dataset.

How it works...

We refer to pandas as pd in *step 1*. In *step 2*, we use read_csv to load the .csv file into a pandas dataframe and call it marketing_data. We also subset the dataframe to include only 11 relevant columns. In *step 3*, we inspect the dataset using head(2) to see the first two rows in the dataset; we also use transpose(T) along with head to transform the rows into columns due to the size of the data (i.e., it has many columns). We use the dtypes attribute of the dataframe to show the data types of all columns. Numeric data has int and float data types while character data has the object data type. We inspect the number of rows and columns using shape, which returns a tuple that displays the number of rows as the first element and the number of columns as the second element.

In *step 4*, we apply the `sort_values` method to sort the `NumStorePurchases` column. Using the `sort_values` method, we sort `NumStorePurchases` in descending order. The method takes two arguments, the dataframe column to be sorted and the sorting order. `false` indicates a sort in descending order while `true` indicates a sort in ascending order.

There's more...

Sorting can be done across multiple columns in `pandas`. We can sort based on multiple columns by supplying columns as a list in the `sort_values` method. The sort will be performed in the order in which the columns are supplied – that is, column 1 first, then column 2 next, and subsequent columns. Also, a sort isn't limited to numerical columns alone; it can be used for columns containing characters.

Categorizing data

When we refer to categorizing data, we are specifically referring to binning, bucketing, or cutting a dataset. Binning involves grouping the numeric values in a dataset into smaller intervals called *bins* or *buckets*. When we bin numerical values, each bin becomes a categorical value. Bins are very useful because they can provide us with insights that may have been difficult to spot if we had worked directly with individual numerical values. Bins don't always have equal intervals; the creation of bins is dependent on our understanding of a dataset.

Binning can also be used to address outliers or reduce the effect of observation errors. Outliers are unusually high or unusually low data points that are far from other data points in our dataset. They typically lead to anomalies in the output of our analysis. Binning can reduce this effect by placing the range of numerical values including the outliers into specific buckets, thereby making the values categorical. A common example of this is when we convert age values into age groups. Outlier ages such as 0 or 150 can fall into a less than age 18 bin and greater than age 80 bin respectively.

In `pandas`, the `cut` method can be used to bin a dataset.

Getting ready

We will work with the full Marketing Campaign data for this recipe.

How to do it...

We will categorize data using the `pandas` library:

1. Import the `pandas` library:

    ```
    import pandas as pd
    ```

2. Load the `.csv` file into a dataframe using `read_csv`. Then, subset the dataframe to include only relevant columns:

```
marketing_data = pd.read_csv("data/marketing_campaign.
csv")


marketing_data = marketing_data[['ID','Year_
Birth', 'Education','Marital_
Status','Income','Kidhome', 'Teenhome', 'Dt_
Customer',                    'Recency','NumStorePurchases',
'NumWebVisitsMonth']]
```

3. Inspect the data. Check the first few rows and use `transpose` (T) to show more information. Also, check the data types as well as the number of columns and rows:

```
marketing_data.head(2).T
                0      1
ID      5524     2174
Year_Birth     1957      1954
Education    Graduation    Graduation
...      ...       ...
NumWebVisitsMonth    7     5


marketing_data.dtypes
ID      int64
Year_Birth     int64
Education    object
...      ...
NumWebVisitsMonth     int64


marketing_data.shape
(2240, 11)
```

4. Categorize the number of store purchases into high, moderate, and low categories:

```
marketing_data['bins'] = pd.cut(x=marketing_
data['NumStorePurchases'], bins=[0,4,8,13],labels =
['Low', 'Moderate', 'High'])
```

5. Inspect the result. Subset for relevant columns:

```
marketing_data[['NumStorePurchases','bins']].head()
     NumStorePurchases      bins
0        4          Low
1        2          Low
2        10         High
3        4          Low
4        6          Moderate
```

We have now categorized our dataset into bins.

How it works...

We refer to pandas as pd in *step 1*. In *step 2*, we use read_csv to load the .csv file into a pandas dataframe and call it marketing_data. We also subset the dataframe to include only 11 relevant columns. In *step 3*, we inspect the dataset using head(2) to see the first two rows in the dataset; we also use transpose(T) along with head to transform the rows into columns due to the size of the data (i.e., it has many columns). We use the dtypes attribute of the dataframe to show the data types of all columns. Numeric data has int and float data types while character data has the object data type. We inspect the number of rows and columns using shape, which returns a tuple that displays the number of rows and columns respectively.

In *step 4*, we categorize the number of store purchases into three categories, namely High, Moderate, and Low. Using the cut method, we cut NumStorePurchases into these three bins and supply the logic for binning within the bin parameter, which is the second parameter. The third parameter is the label parameter. Whenever we manually supply the bin edges in a list as done previously, the bins are typically the number of label categories + 1.

Our bins can be interpreted as 0–4 (low), 5–8 (moderate), and 9–13 (high). In *step 5*, we subset for relevant columns and inspect the result of our binning.

There's more...

For the bin argument, we can also supply the number of bins we require instead of supplying the bin edges manually. This means in the previous steps, we could have supplied the value 3 to the bin parameter and the cut method would have categorized our data into three equally spaced bins. When the value 3 is supplied, the cut method focuses on the equal spacing of the bins, even though the number of records in the bins may be different.

If we are also interested in the distribution of our bins and not just equally spaced bins or user-defined bins, the qcut method in pandas can be used. The qcut method ensures the distribution of data in the bins is equal. It ensures all bins have (roughly) the same number of observations, even though the bin range may vary.

Removing duplicate data

Duplicate data can be very misleading and can lead us to wrong conclusions about patterns and the distribution of our data. Therefore, it is very important to address duplicate data within our dataset before embarking on any analysis. Performing a quick duplicate check is good practice in EDA. When working with tabular datasets, we can identify duplicate values in specific columns or duplicate records (across multiple columns). A good understanding of our dataset and the domain will give us insight into what should be considered a duplicate. In pandas, the drop_duplicates method can help us with handling duplicate values or records within our dataset.

Getting ready

We will work with the full *Marketing Campaign* data for this recipe.

How to do it...

We will remove duplicate data using the pandas library:

1. Import the pandas library:

    ```
    import pandas as pd
    ```

2. Load the .csv file into a dataframe using read_csv. Then, subset the dataframe to include only relevant columns:

    ```
    marketing_data = pd.read_csv("data/marketing_campaign.
    csv")
    marketing_data = marketing_data[['Education','Marital_
    Status','Kidhome', 'Teenhome']]
    ```

3. Inspect the data. Check the first few rows. Also, check the number of columns and rows:

    ```
    marketing_data.head()
             Education    Marital_Status    Kidhome    Teenhome
    0    Graduation    Single      0    0
    1    Graduation    Single      1    1
    2    Graduation    Together    0    0
    3    Graduation    Together    1    0
    ```

```
4     PhD      Married     1     0
```

```
marketing_data.shape
(2240, 4)
```

4. Remove duplicates across the four columns in our dataset:

```
marketing_data_duplicate = marketing_data.drop_
duplicates()
```

5. Inspect the result:

```
marketing_data_duplicate.head()
       Education      Marital_Status      Kidhome      Teenhome
0      Graduation     Single     0     0
1      Graduation     Single     1     1
2      Graduation     Together   0     0
3      Graduation     Together   1     0
4      PhD      Married     1     0
```

```
marketing_data_duplicate.shape
(135,4)
```

We have now removed duplicates from our dataset.

How it works...

We refer to pandas as pd in *step 1*. In *step 2*, we use read_csv to load the .csv file into a pandas dataframe and call it marketing_data. We also subset the dataframe to include only four relevant columns. In *step 3*, we inspect the dataset using head() to see the first five rows in the dataset. Using the shape method, we get a sense of the number of rows and columns from the tuple respectively.

In *step 4*, we use the drop_duplicates method to remove duplicate rows that appear in the four columns of our dataset. We save the result in the marketing_data_duplicate variable. In *step 5*, we inspect the result using the head method to see the first five rows. We also leverage the shape method to inspect the number of rows and columns. We can see that the rows have decreased significantly from our original shape.

There's more...

The `drop_duplicates` method gives some flexibility around dropping duplicates based on a subset of columns. By supplying the list of the subset columns as the first argument, we can drop all rows that contain duplicates based on those subset columns. This is useful when we have several columns and only a few key columns contain duplicate information. Also, it allows us to keep instances of duplicates, using the `keep` parameter. With the `keep` parameter, we can specify whether we want to keep the "first" or "last" instance or drop all instances of the duplicate information. By default, the method keeps the first instance.

Dropping data rows and columns

When working with tabular data, we may have reason to drop some rows or columns within our dataset. Sometimes, we may need to drop columns or rows either because they are erroneous or irrelevant. In `pandas`, we have the flexibility to drop a single row/column or multiple rows/columns. We can use the `drop` method to achieve this.

Getting ready

We will work with the full *Marketing Campaign* data for this recipe.

How to do it...

We will drop rows and columns using the `pandas` library:

1. Import the `pandas` library:

    ```
    import pandas as pd
    ```

2. Load the `.csv` file into a dataframe using `read_csv`. Then, subset the dataframe to include only relevant columns:

    ```
    marketing_data = pd.read_csv("data/marketing_campaign.
    csv")
    marketing_data = marketing_data[['ID', 'Year_Birth',
    'Kidhome', 'Teenhome']]
    ```

3. Inspect the data. Check the first few rows. Check the number of columns and rows:

    ```
    marketing_data.head()
        ID    Year_Birth    Education    Marital_Status
    0   5524    1957        Graduation    Single
    1   2174    1954        Graduation    Single
    ```

```
2     4141    1965    Graduation    Together
3     6182    1984    Graduation    Together
4     5324    1981    PhD     Married
```

```
marketing_data.shape
(5, 4)
```

4. Delete a specified row at index value 1:

```
marketing_data.drop(labels=[1], axis=0)
        ID    Year_Birth    Education    Marital_Status
0     5524    1957    Graduation    Single
2     4141    1965    Graduation    Together
3     6182    1984    Graduation    Together
4     5324    1981    PhD     Married
```

5. Delete a single column:

```
marketing_data.drop(labels=['Year_Birth'], axis=1)
        ID    Education    Marital_Status
0     5524    Graduation    Single
1     2174    Graduation    Single
2     4141    Graduation    Together
3     6182    Graduation    Together
4     5324    PhD     Married
```

Good job! We have dropped rows and columns from our dataset.

How it works...

We refer to pandas as pd in *step 1*. In *step 2*, we use read_csv to load the .csv file into a pandas dataframe and call it marketing_data. We also subset the dataframe to include only four relevant columns. In *step 3*, we inspect the dataset using head() to see the first five rows in the dataset. Using the shape method, we get a sense of the number of rows and columns from the tuple respectively.

In *step 4*, we use the drop method to delete a specified row at index value 1 and view the result, which shows the row at index 1 has been removed. The drop method takes a list of indices as the first argument and an axis value as the second. The axis value determines whether the drop operation will be performed on a row or column. A value of 0 is used for rows while 1 is used for columns.

In *step 5*, we use the `drop` method to delete a specified column and view the result, which shows the specific column has been removed. To drop columns, we need to specify the name of the column and provide the `axis` value of 1.

There's more...

We can drop multiple rows or columns using the `drop` method. To achieve this, we need to specify all the row indices or column names in a list and provide the respective axis value of 0 or 1 for rows and columns respectively.

Replacing data

Replacing values in rows or columns is a common practice when working with tabular data. There are many reasons why we may need to replace specific values within a dataset. Python provides the flexibility to replace single values or multiple values within our dataset. We can use the `replace` method to achieve this.

Getting ready

We will work with the *Marketing Campaign* data again for this recipe.

How to do it...

We will remove duplicate data using the `pandas` library:

1. Import the `pandas` library:

    ```
    import pandas as pd
    ```

2. Load the `.csv` file into a dataframe using `read_csv`. Then, subset the dataframe to include only relevant columns:

    ```
    marketing_data = pd.read_csv("data/marketing_campaign.
    csv")
    marketing_data = marketing_data[['ID', 'Year_Birth',
    'Kidhome', 'Teenhome']]
    ```

3. Inspect the data. Check the first few rows, and check the number of columns and rows:

	ID	Year_Birth	Kidhome	Teenhome
0	5524	1957	0	0
1	2174	1954	1	1
2	4141	1965	0	0

```
3     6182     1984     1     0
4     5324     1981     1     0

marketing_data.shape
(2240, 4)
```

4. Replace the values in Teenhome with has teen and has no teen:

```
marketing_data['Teenhome_replaced'] = marketing_
data['Teenhome'].replace([0,1,2],['has no teen','has
teen','has teen'])
```

5. Inspect the output:

```
marketing_data[['Teenhome','Teenhome_replaced']].head()
       Teenhome     Teenhome_replaced
0     0        has no teen
1     1        has teen
2     0        has no teen
3     0        has no teen
4     0        has no teen
```

Great! We just replaced values in our dataset.

How it works...

We refer to pandas as *pd* in *step 1*. In *step 2*, we use read_csv to load the .csv file into a pandas dataframe and call it marketing_data. We also subset the dataframe to include only four relevant columns. In *step 3*, we inspect the dataset using head() to see the first five rows in the dataset. Using the shape method, we get a sense of the number of rows and columns.

In *step 4*, we use the replace method to replace values within the Teenhome column. The first argument of the method is a list of the existing values that we want to replace, while the second argument contains a list of the values we want to replace it with. It is important to note that the lists for both arguments must be the same length.

In *step 5*, we inspect the result.

There's more...

In some cases, we may need to replace a group of values that have complex patterns that cannot be explicitly stated. An example could be certain phone numbers or email addresses. In such cases, the `replace` method gives us the ability to use regex for pattern matching and replacement. **Regex** is short for **regular expressions**, and it is used for pattern matching.

See also

- You can check out this great resource by *Data to Fish* on replacing data in `pandas`: `https://datatofish.com/replace-values-pandas-dataframe/`
- Here is an insightful resource by *GeeksforGeeks* on regex in the `replace` method in pandas: `https://www.geeksforgeeks.org/replace-values-in-pandas-dataframe-using-regex/`
- Here is another article by *W3Schools* that highlights common regex patterns in Python: `https://www.w3schools.com/python/python_regex.asp`

Changing a data format

When analyzing or exploring data, the type of analysis we perform on our data is highly dependent on the data formats or data types within our dataset. Typically, numerical data requires specific analytical techniques, while categorical data requires other analytical techniques. Hence, it is important that data types are properly captured before analysis commences.

In `pandas`, the `dtypes` attribute helps us to inspect the data types within our dataset, while the `astype` attribute helps us to convert our dataset between various data types.

Getting ready

We will work with the *Marketing Campaign* data again for this recipe.

How to do it...

We will change the format of our data using the `pandas` library:

1. Import the `pandas` library:

    ```
    import pandas as pd
    ```

2. Load the `.csv` file into a dataframe using `read_csv`. Then, subset the dataframe to include only relevant columns:

```
marketing_data = pd.read_csv("data/marketing_campaign.
csv")
marketing_data = marketing_data[['ID', 'Year_
Birth','Marital_Status','Income']]
```

3. Inspect the data. Check the first few rows. Check the number of columns and rows:

```
        ID    Year_Birth     Marital_
   Status    Income    Income_changed
0    5524    1957    Single    58138.0    58138
1    2174    1954    Single    46344.0    46344
2    4141    1965    Together    71613.0    71613
3    6182    1984    Together    26646.0    26646
4    5324    1981    Married    58293.0    58293


marketing_data.shape
(2240, 5)
```

4. Fill NAs in the `Income` column:

```
marketing_data['Income'] = marketing_data['Income'].
fillna(0)
```

5. Change the data type of the `Income` column from float to int:

```
marketing_data['Income_changed'] = marketing_
data['Income'].astype(int)
```

6. Inspect the output using the `head` method and `dtypes` attribute:

```
marketing_data[['Income','Income_changed']].head()
      Income    Income_changed
0    58138.0    58138
1    46344.0    46344
2    71613.0    71613
3    26646.0    26646
4    58293.0    58293


marketing_data[['Income','Income_changed']].dtypes
```

```
        0
Income       float64
Income_changed      int32
```

Now we have changed the format of our dataset.

How it works...

We refer to pandas as pd in *step 1*. In *step 2*, we use `read_csv` to load the csv file into a pandas dataframe and call it `marketing_data`. We also subset the dataframe to include only four relevant columns. In *step 3*, we inspect the dataset using `head()` to see the first five rows in the dataset. Using the `shape` method, we get a sense of the number of rows and columns.

In *step 4*, we fill NaN values with zeros using the `fillna` method. This is an important step before we can change data types in pandas. We have provided the `fillna` method with the `just` argument, which is the value to replace NaN values with. In *step 5*, we change the data type of the `Income` column from float to int using the `astype` method. We supply the data type we wish to convert to as the argument for the method.

In *step 6*, we subset the dataframe and inspect the result.

There's more...

When converting data types, we may encounter errors in conversion. The `astype` method gives us options through the `errors` parameter to raise or ignore errors. By default, the method raises errors; however, we can ignore errors so that the method returns the original values for each error identified.

See also

Here is a great article by *PB Python* that provides more details on converting data types in pandas: `https://pbpython.com/pandas_dtypes.html`.

Dealing with missing values

Dealing with missing values is a common problem we will typically face when analyzing data. A missing value is a value within a field or variable that is not present, even though it is expected to be. There are several reasons why this could have happened, but a common reason is that the data value wasn't provided at the point of data collection. As we explore and analyze data, missing values can easily lead to inaccurate or biased conclusions; therefore, they need to be taken care of. Missing values are typically represented by blank spaces, but in pandas, they are represented by NaN.

Several techniques can be used to deal with missing values. In this recipe, we will focus on dropping missing values using the `dropna` method in pandas.

Getting ready

We will work with the full *Marketing Campaign* data in this recipe.

How to do it...

We will drop rows and columns using the `pandas` library:

1. Import the `pandas` library:

    ```
    import pandas as pd
    ```

2. Load the `.csv` into a dataframe using `read_csv`. Then, subset the dataframe to include only relevant columns:

    ```
    marketing_data = pd.read_csv("data/marketing_campaign.
    csv")
    marketing_data = marketing_data[['ID', 'Year_Birth',
    'Education','Income']]
    ```

3. Inspect the data. Check the first few rows, and check the number of columns and rows:

    ```
    marketing_data.head()
         ID     Year_Birth          Education      Income
    0    5524      1957             Graduation     58138.0
    1    2174      1954             Graduation     46344.0
    2    4141      1965             Graduation     71613.0
    3    6182      1984             Graduation     26646.0
    4    5324      1981             PhD            58293.0

    marketing_data.shape
    (2240, 4)
    ```

4. Check for missing values using the `isnull` and `sum` methods:

    ```
    marketing_data.isnull().sum()
    ID               0
    Year_Birth       0
    Education        0
    Income          24
    ```

5.   Drop missing values using the `dropna` method:

```
marketing_data_withoutna = marketing_data.dropna(how =
'any')
marketing_data_withoutna.shape
(2216, 4)
```

Good job! We have dropped missing values from our dataset.

How it works...

In *step 1*, we import `pandas` and refer to it as pd. In *step 2*, we use `read_csv` to load the `.csv` file into a `pandas` dataframe and call it `marketing_data`. We also subset the dataframe to include only four relevant columns. In *step 3*, we inspect the dataset using the `head` and `shape` methods.

In *step 4*, we use the `isnull` and `sum` methods to check for missing values. These methods give us the number of rows with missing values within each column in our dataset. Columns with zero have no rows with missing values.

In *step 5*, we use the `dropna` method to drop missing values. For the `how` parameter, we supply `'any'` as the value to indicate that we want to drop rows that have missing values in any of the columns. An alternative value to use is `'all'`, which ensures all the columns of a row have missing values before dropping the row. We then check the number of rows and columns using the `shape` method. We can see that the final dataset has 24 fewer rows.

There's more...

As highlighted previously, there are several reasons why a value may be missing in our dataset. Understanding the reason can point us to optimal solutions to resolve this problem. Missing values shouldn't be addressed with a one-size-fits-all approach. *Chapter 9* dives into the details of how to optimally deal with missing values and outliers by providing several techniques.

See also

You can check out a detailed approach to dealing with missing values and outliers in *Chapter 9, Dealing with Outliers and Missing Values.*

Visualizing Data in Python

Visualizing data is a critical component of **Exploratory Data Analysis (EDA)**. It helps us understand relationships, patterns, and hidden trends within our data much better. With the right charts or visuals, trends within large and complex datasets can be easily interpreted, and hidden patterns or outliers can be easily identified. In Python, data can be visualized using a wide array of libraries, and in this chapter, we will explore the most common data visualization libraries in Python.

In this chapter, we will cover the following key topics:

- Preparing for visualization
- Visualizing data in Matplotlib
- Visualizing data in Seaborn
- Visualizing data in GGPLOT
- Visualizing data in Bokeh

Technical requirements

We will leverage the `pandas`, `matplotlib`, `seaborn`, `plotnine`, and `bokeh` libraries in Python for this chapter. The code and notebooks for this chapter are available on GitHub at `https://github.com/PacktPublishing/Exploratory-Data-Analysis-with-Python-Cookbook`.

Preparing for visualization

Before visualizing our data, it is important to get a glimpse of what the data looks like. This step basically involves inspecting our data to get a sense of the shape, data types, and type of information. Without this critical step, we may end up using the wrong visuals for analyzing our data. Visualizing data is never one size fits all because different charts and visuals require different data types and numbers of variables. This must always be factored in when visualizing data. Also, we may be required to transform our data before EDA; inspecting our data helps us identify whether transformation is required before proceeding to EDA. Lastly, this step helps us to identify whether there are additional variables that can be created from transforming or combining existing variables (feature engineering).

In pandas, the head, dtypes, and shape attributes are good ways to get a glimpse of our data.

Getting ready

We will work with one dataset in this chapter: **Amsterdam House Prices Data** from Kaggle.

Start by creating a folder for this chapter and then add a new Python script or Jupyter Notebook file to that folder. Create a data subfolder and then place the HousingPricesData.csv file in that subfolder. You can also retrieve all the files from the GitHub repository.

> **Note**
>
> Kaggle provides Amsterdam House Prices Data for public use at https://www.kaggle.com/ datasets/thomasnibb/amsterdam-house-price-prediction. In this chapter, we will use the full dataset for the different recipes. The data is also available in the repository.

How to do it...

We will learn how to prepare for EDA using the pandas library:

1. Import the pandas library:

    ```
    import pandas as pd
    ```

2. Load the .csv into a dataframe using read_csv. Then subset the dataframe to include only relevant columns:

    ```
    houseprices_data = pd.read_csv("data/HousingPricesData.csv")


    houseprices_data = houseprices_data[['Zip', 'Price', 'Area',
    'Room']]
    ```

3. Inspect the data. Check the first five rows using the head method. Also, check the data types as well as the number of columns and rows:

    ```
    houseprices_data.head()
         Zip         Price      Area    Room
    0    1091 CR     685000.0   64      3
    1    1059 EL     475000.0   60      3
    2    1097 SM     850000.0   109     4
    3    1060 TH     580000.0   128     6
    4    1036 KN     720000.0   138     5
    ```

```
houseprices_data.shape
(924,5)

houseprices_data.dtypes
Zip      object
Price    float64
Area     int64
Room     int64
```

4. Create a `PriceperSqm` variable based on the `Price` and `Area` variables:

```
houseprices_data['PriceperSqm'] = houseprices_data['Price']/
houseprices_data['Area']

houseprices_data.head()
     Zip        Price      Area    Room     PriceperSqm
0    1091 CR    685000.0   64      3        10703.125
1    1059 EL    475000.0   60      3        7916.6667
2    1097 SM    850000.0   109     4        7798.1651
3    1060 TH    580000.0   128     6        4531.25
4    1036 KN    720000.0   138     5        5217.3913
```

That's all. Now we have inspected our data in preparation for visualization, and we have also created a new variable.

How it works...

In this recipe, we use the `pandas` library to inspect and get a quick glimpse of our dataset. We import `pandas` and refer to it as `pd` in step 1. In step 2, we use `read_csv` to load the `.csv` file into a `pandas` dataframe and call it `houseprices_data`. We subset the dataframe to include only four relevant columns. In step 3, we get a quick glimpse of our data by inspecting the first five rows using the head method. We also get a sense of the dataframe shape (number of rows and columns) and the data types using the `shape` and `dtypes` attributes, respectively.

In step 4, we create the `PriceperSqm` variable based on domain knowledge. This metric is common in real estate, and it is created by dividing the house price by the area (in square meters). This new variable will be useful during EDA.

There's more...

Sometimes we may be required to change the data types of specific columns before proceeding to EDA. This is another important reason why inspecting our data is critical before commencing EDA.

Visualizing data in Matplotlib

matplotlib is a data visualization library for creating static and interactive visualizations in Python. It contains a wide variety of visual options, such as line charts, bar charts, histograms, and many more. It is built on NumPy arrays. matplotlib is a low-level API that provides flexibility for both simple and complex visuals. However, this also means that achieving simple tasks can be quite cumbersome.

The `pyplot` module within the matplotlib library handles visualization needs. Some important concepts we will typically encounter while using the `matplotlib` library include the following:

- **Figure**: This is the frame within which graphs are plotted. Simply put, the figure is where the plotting is done.
- **Axes**: These are the horizontal and vertical lines (*x* and *y* axes) that provide the border for the graph and act as a reference for measurement.
- **Ticks**: These are the small lines that help us demarcate our axes.
- **Title**: This is the title of our graph within the figure.
- **Labels**: These are the labels for our ticks along the axes.
- **Legend**: This provides additional information about the chart to aid the correct interpretation.

Most of the preceding concepts also apply to various visualization libraries in Python.

Now let's explore `matplotlib` using examples.

Getting ready

We will continue working with Amsterdam House Prices Data from Kaggle.

We will also use the `matplotlib` library for this recipe. You can install `matplotlib` with `pip` using the following command:

```
pip install matplotlib
```

How to do it...

We will learn how to visualize data using `matplotlib`:

1. Import the `pandas` and `matplotlib` libraries:

    ```
    import pandas as pd
    import matplotlib.pyplot as plt
    ```

2. Load the `.csv` into a dataframe using `read_csv`. Then subset the dataframe to include only relevant columns:

```
houseprices_data = pd.read_csv("data/HousingPricesData.csv")


houseprices_data = houseprices_data[['Zip', 'Price', 'Area',
'Room']]
```

3. Inspect the data. Get a glimpse of the first five rows using the head method. Also, get a sense of the data types as well as the number of columns and rows:

```
houseprices_data.head()
     Zip         Price     Area    Room
0    1091 CR     685000.0   64     3
1    1059 EL     475000.0   60     3
2    1097 SM     850000.0   109    4
3    1060 TH     580000.0   128    6
4    1036 KN     720000.0   138    5


houseprices_data.shape
(924,5)


houseprices_data.dtypes
Zip       object
Price       float64
Area        int64
Room        int64
```

4. Create a price per sqm variable based on the `Price` and `Area` variables:

```
houseprices_data['PriceperSqm'] = houseprices_data['Price']/
houseprices_data['Area']


houseprices_data.head()
     Zip         Price     Area    Room    PriceperSqm
0    1091 CR     685000.0   64     3       10703.125
1    1059 EL     475000.0   60     3       7916.6667
2    1097 SM     850000.0   109    4       7798.1651
3    1060 TH     580000.0   128    6       4531.25
4    1036 KN     720000.0   138    5       5217.3913
```

5. Sort the dataframe based on the house prices and inspect the output:

```
houseprices_sorted = houseprices_data.sort_values('Price',
ascending = False)
houseprices_sorted.head()
```

```
        Zip          Price      Area   Room      PriceperSqm
195    1017 EL     5950000.0    394    10       15101.5228
837    1075 AH     5850000.0    480    14       12187.5
305    1016 AE     4900000.0    623    13       7865.1685
103    1017 ZP     4550000.0    497    13       9154.9295
179    1012 JS     4495000.0    178     5       25252.8089
```

6. Plot a bar chart in `matplotlib` with basic details:

```
plt.figure(figsize= (12,6))

x = houseprices_sorted['Zip'][0:10]
y = houseprices_sorted['Price'][0:10]
plt.bar(x,y)
plt.show()
```

This results in the following output:

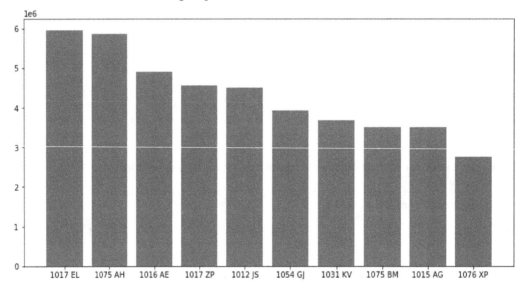

Figure 3.1: A matplotlib bar chart with basic details

7. Plot a bar chart in `matplotlib` with additional informative details such as the title, x label, and y label:

```
plt.figure(figsize= (12,6))
plt.bar(x,y)
plt.title('Top 10 Areas with the highest house prices',
fontsize=15)
```

```
plt.xlabel('Zip code', fontsize = 12)
plt.xticks(fontsize=10)
plt.ylabel('House prices in millions', fontsize=12)
plt.yticks(fontsize=10)
plt.show()
```

This results in the following output:

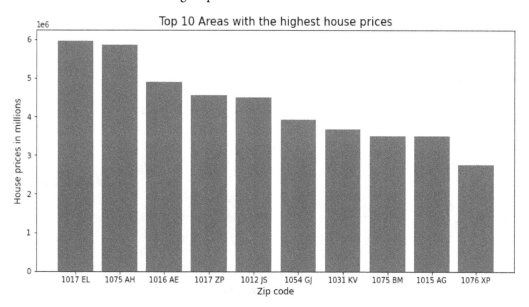

Figure 3.2: A matplotlib bar chart with additional details

8. Create subplots in `matplotlib` to view multiple perspectives at once:

```
fig, ax = plt.subplots(figsize=(40,18))

x = houseprices_sorted['Zip'][0:10]
y = houseprices_sorted['Price'][0:10]
y1 = houseprices_sorted['PriceperSqm'][0:10]

plt.subplot(1,2,1)
plt.bar(x,y)
plt.xticks(fontsize=17)
plt.ylabel('House prices in millions', fontsize=25)
plt.yticks(fontsize=20)
plt.title('Top 10 Areas with the highest house prices',
fontsize=25)

plt.subplot(1,2,2)
plt.bar(x,y1)
```

```
plt.xticks(fontsize=17)
plt.ylabel('House prices per sqm', fontsize=25)
plt.yticks(fontsize=20)
plt.title('Top 10 Areas with the highest house prices per sqm',
fontsize=25)
plt.show()
```

This results in the following output:

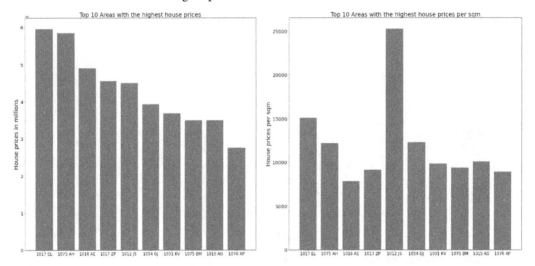

Figure 3.3: The matplotlib subplots

Great. Now we are done exploring matplotlib.

How it works...

In step 1, we import the pandas library and refer to it as pd. We also import the pyplot module from the matplotlib library. We refer to it as plt; this module will serve all our visualization needs. In step 2, we use read_csv to load the .csv file into a pandas dataframe, and then we call it houseprices_data.

In step 3, we inspect the dataset using the head method to get a sense of the first five rows in the dataset; we also use the dtypes and shape attributes to get the data types and the number of rows and columns, respectively.

In step 4, we create the price per square meter variable. In step 5, we sort the dataframe based on the Price column using the sort_values method. This will be useful since we plan to visualize the top 10 areas based on house price. In step 6, we plot a bar chart using matplotlib. We start by using the figure function in pyplot to define the size of the frame within which plotting will be done. Using the figsize parameter, we specify the width and the height respectively. Next, we define the values in the *x* axis by extracting the zip codes of the first 10 records in our sorted house prices dataframe; this will give us the zip codes of the top 10 most expensive houses. We also define the values in the *y* axis by extracting the price of the first 10 records of the same dataframe. Using the `bar` function, we then plot our bar chart, and using the `show` function, we can display the bar chart we plotted.

In step 7, we provide additional details to make our bar chart informative. The `title` function provides a title and the `xlabel` and `ylabel` functions provide labels for our *x* axis and *y* axis, respectively. These functions take in the text as the first argument and the size of the text as the second argument. The `xticks` and `yticks` functions specify the size of our *x*- and *y-axis* values, respectively.

In step 8, we create subplots using the `subplots` function. The function returns a tuple with two elements. The first element is `figure`, which is the frame of the plot, while the second represents the axes, which are the canvas we draw on. We assign the two elements in the tuple to two variables: `fig` and `ax`. We then specify the position of our first subplot and provide the details of the plot. The function takes in three arguments, number of rows (1), number of columns (2), and index (1). Next, we specify the position of the second plot, number of rows (1), number of columns (2), and index (2), then provide the details.

Our first subplot highlights the top 10 areas based on house prices, while our second subplot highlights the same areas but based on price per sqm. Analyzing the two plots side by side highlights a hidden pattern that we may have missed without EDA; for example, the most expensive area by price doesn't necessarily have the highest price per sqm.

There's more...

The `matplotlib` library supports several types of charts and customization options. The low-level API provides flexibility for several visual requirements, both simple and complex. Some other charts supported in `matplotlib` include histograms, boxplots, violin plots, pie charts, and so on.

See also

- Check out the `matplotlib` documentation: `https://matplotlib.org/stable/users/index.html`

- Here is an insightful article in *Towards Data Science* by Badreesh Shetty: `https://towardsdatascience.com/data-visualization-using-matplotlib-16f1aae5ce70`

Visualizing data in Seaborn

Seaborn is another common Python data visualization library. It is based on `matplotlib` and integrates well with `pandas` data structures. Seaborn is primarily used for making statistical graphics, and this makes it a very good candidate for performing EDA. It uses matplotlib to draw its charts; however, this is done behind the scenes. Unlike matplotlib, `seaborn`'s high-level API makes it easier and faster to use. As mentioned earlier, common tasks can sometimes be cumbersome in matplotlib and take several lines of code. Even though matplotlib is highly customizable, the settings can sometimes be hard to tweak. However, seaborn provides settings that are easier to tweak and understand.

Many of the important terms considered under the previous recipe also apply to seaborn, such as axes, ticks, legends, titles, labels, and so on. We will explore seaborn through some examples.

Getting ready

We will continue working with Amsterdam House Prices Data from Kaggle.

We will also use the matplotlib and seaborn libraries for this recipe. You can install seaborn with pip using the following command:

```
pip install seaborn
```

How to do it...

We will learn how to visualize data using seaborn:

1. Import the pandas, matplotlib, and seaborn libraries:

    ```
    import pandas as pd
    import matplotlib.pyplot as plt
    import seaborn as sns
    ```

2. Load the .csv into a dataframe using read_csv. Then subset the dataframe to include only relevant columns:

    ```
    houseprices_data = pd.read_csv("data/HousingPricesData.csv")

    houseprices_data = houseprices_data[['Zip', 'Price', 'Area',
    'Room']]
    ```

3. Inspect the data. Get a glimpse of the first five rows using the head method. Also, get a sense of the data types as well as the number of columns and rows:

    ```
    houseprices_data.head()
         Zip         Price      Area      Room
    0    1091 CR     685000.0   64        3
    1    1059 EL     475000.0   60        3
    2    1097 SM     850000.0   109       4
    3    1060 TH     580000.0   128       6
    4    1036 KN     720000.0   138       5
    ```

```
houseprices_data.shape
(924,5)

houseprices_data.dtypes
Zip      object
Price    float64
Area     int64
Room     int64
```

4. Create a price per sqm variable based on the Price and Area variables:

```
houseprices_data['PriceperSqm'] = houseprices_data['Price']/
houseprices_data['Area']

houseprices_data.head()
      Zip         Price       Area    Room     PriceperSqm
0     1091 CR     685000.0    64      3        10703.125
1     1059 EL     475000.0    60      3        7916.6667
2     1097 SM     850000.0    109     4        7798.1651
3     1060 TH     580000.0    128     6        4531.25
4     1036 KN     720000.0    138     5        5217.3913
```

5. Sort the dataframe based on the house prices and inspect the output:

```
houseprices_sorted = houseprices_data.sort_values('Price',
ascending = False)
houseprices_sorted.head()
        Zip         Price        Area    Room     PriceperSqm
195     1017 EL     5950000.0    394     10       15101.5228
837     1075 AH     5850000.0    480     14       12187.5
305     1016 AE     4900000.0    623     13       7865.1685
103     1017 ZP     4550000.0    497     13       9154.9295
179     1012 JS     4495000.0    178     5        25252.8089
```

6. Plot a bar chart in seaborn with basic details:

```
plt.figure(figsize= (12,6))
data = houseprices_sorted[0:10]
sns.barplot(data= data, x= 'Zip',y = 'Price')
```

This results in the following output:

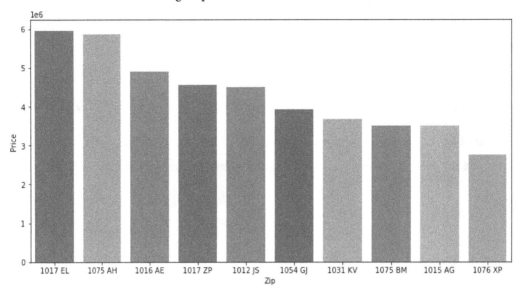

Figure 3.4: A seaborn bar chart with basic details

7. Plot a bar chart in `seaborn` with additional informative details such as the title, x label, and y label:

```
plt.figure(figsize= (12,6))
data = houseprices_sorted[0:10]

ax = sns.barplot(data= data, x= 'Zip',y = 'Price')
ax.set_xlabel('Zip code',fontsize = 15)
ax.set_ylabel('House prices in millions', fontsize = 15)
ax.set_title('Top 10 Areas with the highest house prices',
fontsize= 20)
```

This results in the following output:

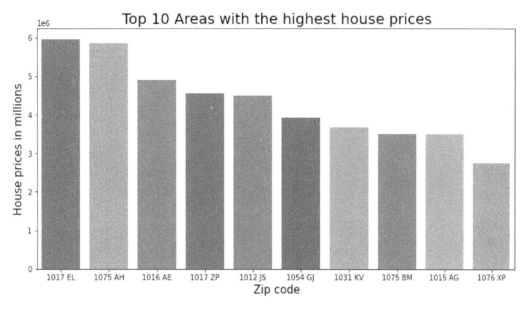

Figure 3.5: A seaborn bar chart with additional details

8. Create subplots in seaborn to view multiple perspectives at once:

```
fig, ax = plt.subplots(1, 2,figsize=(40,18))

data = houseprices_sorted[0:10]

sns.set(font_scale = 3)
ax1 = sns.barplot(data= data, x= 'Zip',y = 'Price', ax = ax[0])
ax1.set_xlabel('Zip code')
ax1.set_ylabel('House prices in millions')
ax1.set_title('Top 10 Areas with the highest house prices')


ax2 = sns.barplot(data= data, x= 'Zip',y = 'PriceperSqm',
ax=ax[1])
ax2.set_xlabel('Zip code')
ax2.set_ylabel('House prices per sqm')
ax2.set_title('Top 10 Areas with the highest price per sqm')
```

This results in the following output:

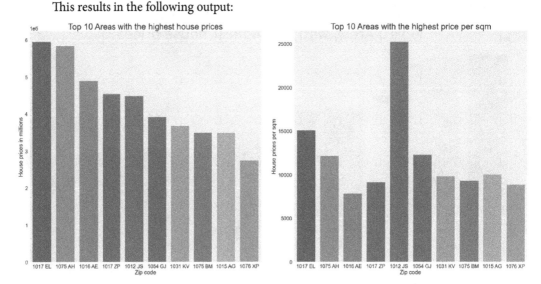

Figure 3.6: The seaborn subplots

Awesome, we have plotted a bar chart in `seaborn`.

How it works...

In step 1, we import the `pandas` library and refer to it as `pd`. We also import the `pyplot` module from the `matplotlib` library and import the `seaborn` library. We refer to `pyplot` as `plt` and `seaborn` as `sns`. In step 2, we use `read_csv` to load the `.csv` file into a `pandas` dataframe, and then we call the `houseprices_data` dataframe.

In step 3, we inspect the dataset using the `head` method to get a sense of the first five rows in the dataset; we also use the `dtypes` and `shape` attributes to get the data types and the number of rows and columns, respectively.

In step 4, we create the price per square meter variable. In step 5, we sort the dataframe based on the `Price` column using the `sort_values` method. This will be useful since we plan to visualize the top 10 areas based on house price. In step 6, we plot a bar chart using `seaborn`. We use the `figure` function in `pyplot` to define the size of our plotting frame. Next, we define our data, which is the first 10 records of our sorted dataset. Using the `barplot` function, we then plot our bar chart.

In step 7, we provide additional details to make our bar chart informative. First, we assign our bar plot to a variable and then use the `seaborn` methods to provide additional chart information. The `set_title` method provides a title, and the `set_xlabel` and `set_ylabel` methods provide labels for our *x* axis and *y* axis, respectively. These methods take in the text as the first argument and the size of the text as our second argument.

In step 8, we create subplots using the `subplots` function from `matplotlib`. We assign the `subplots` function output to two variables (`fig` and `ax`) because it typically outputs a tuple with two elements. In the `subplot` function, we specify the structure of the subplot (one row and two columns) and the figure size. We then specify the position of our first subplot using the `ax` parameter in our `barplot` method. The `ax[0]` value specifies that the first plot goes to the first axis. We do the same for the second subplot. The `set` method helps us set the scale of our subplot's labels and text.

There's more...

In seaborn, we can use the matplotlib functions to provide additional information to our charts. This means functions such as `title`, `xlabel`, `ylabel`, `xticks`, and `yticks` can also be used in `seaborn`.

See also

Check out the `seaborn` user guide to explore the library in more detail: `https://seaborn.pydata.org/tutorial.html`

Visualizing data in GGPLOT

`ggplot` is an open source data visualization library originally built within the programming language R. Over the past few years, it has gained significant popularity. It is an implementation of the grammar of graphics, which is a high-level framework for creating plots in a consistent way. `ggplot` also has a Python implementation called `plotnine`.

The grammar of graphics consists of seven components that abstract the low-level details and allow you to focus on building aesthetically appealing visualizations. The components include data, aesthetics, geometric objects, facets, statistical transformations, coordinates, and themes. These components are described here:

- **Data**: This refers to the data we plan to visualize.

- **Aesthetics**: This refers to the variables we want to plot, that is, a single variable (*x* variable) or multiple variables (*x* and *y* variables).

- **Geometric object**: This refers to the graph we plan to use. An example could be a histogram or bar plot.

- **Facets**: This helps us break our data into subsets and visualize these subsets across multiple plots arranged next to each other.

- **Statistical transformations**: This refers to transformations such as summary statistics, for example, mean, median, and so on, to be performed on the data.

- **Coordinates**: This refers to the coordinate options available. This is a Cartesian coordinate system by default.

- **Themes**: The themes provide appealing design options on how to best visualize data. This includes the background color, legend, and so on.

The first three components are compulsory, as we cannot have a plot without them. The other four components are typically optional.

We will explore ggplot in Python (the plotnine library) through some examples.

Getting ready

We will continue working with Amsterdam House Prices Data from Kaggle.

We will also use the plotnine library for this recipe. In Python, ggplot is implemented as a module within plotnine. You can install plotnine with pip using the following command:

```
pip install plotnine
```

How to do it...

We will learn how to visualize data using ggplot in plotnine:

1. Import the pandas and plotnine libraries:

    ```
    import pandas as pd
    from plotnine import *
    ```

2. Load the .csv into a dataframe using read_csv. Then subset the dataframe to include only relevant columns:

    ```
    houseprices_data = pd.read_csv("data/HousingPricesData.csv")

    houseprices_data = houseprices_data[['Zip', 'Price', 'Area',
    'Room']]
    ```

3. Inspect the data. Get a glimpse of the first five rows using the head method. Also, get a sense of the data types as well as the number of columns and rows:

    ```
    houseprices_data.head()
        Zip          Price      Area      Room
    0   1091 CR      685000.0   64        3
    1   1059 EL      475000.0   60        3
    2   1097 SM      850000.0   109       4
    3   1060 TH      580000.0   128       6
    4   1036 KN      720000.0   138       5
    ```

```
houseprices_data.shape
(924,5)

houseprices_data.dtypes
Zip     object
Price   float64
Area    int64
Room    int64
```

4. Create a price per sqm variable based on the `Price` and `Area` variables:

```
houseprices_data['PriceperSqm'] = houseprices_data['Price']/
houseprices_data['Area']

houseprices_data.head()
     Zip          Price      Area    Room     PriceperSqm
0    1091 CR      685000.0   64      3        10703.125
1    1059 EL      475000.0   60      3        7916.6667
2    1097 SM      850000.0   109     4        7798.1651
3    1060 TH      580000.0   128     6        4531.25
4    1036 KN      720000.0   138     5        5217.3913
```

5. Sort the dataframe based on the house prices and inspect the output:

```
houseprices_sorted = houseprices_data.sort_values('Price',
ascending = False)
houseprices_sorted.head()
     Zip          Price      Area   Room     PriceperSqm
195    1017 EL    5950000.0   394    10      15101.5228
837    1075 AH    5850000.0   480    14      12187.5
305    1016 AE    4900000.0   623    13      7865.1685
103    1017 ZP    4550000.0   497    13      9154.9295
179    1012 JS    4495000.0   178    5       25252.8089
```

6. Plot a bar chart in `ggplot` with basic details:

```
chart_data = houseprices_sorted[0:10]

ggplot(chart_data,aes(x='Zip',y = 'Price')) + geom_bar(stat =
'identity') \
+ scale_x_discrete(limits=chart_data['Zip'].tolist()) +
theme(figure_size=(16, 8))
```

This results in the following output:

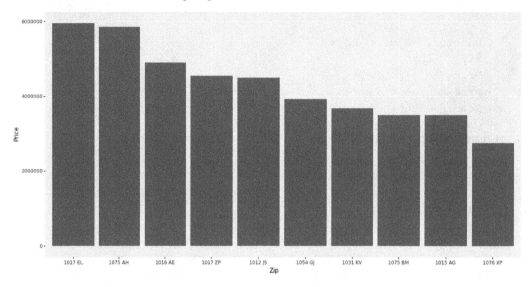

Figure 3.7: A GGPLOT bar chart with basic details

7. Plot a bar chart in `ggplot` with additional informative details such as the title, x label, and y label:

```
ggplot(chart_data,aes(x='Zip',y = 'Price')) + geom_bar(stat =
'identity') \
+ scale_x_discrete(limits=chart_data['Zip'].tolist()) \
+ labs(y='House prices in millions', x='Zip code', title='Top 10
Areas with the highest house prices') \
+ theme(figure_size=(16, 8),
        axis_title=element_text(face='bold',size =12),
        axis_text=element_text(face='italic',size=8),
        plot_title=element_text(face='bold',size=12))
```

This results in the following output:

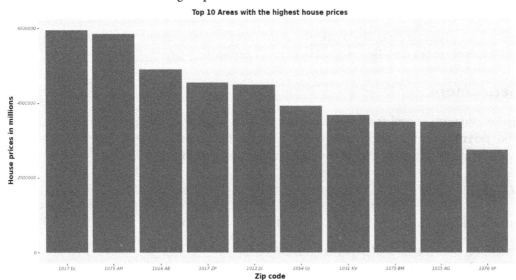

Figure 3.8: A ggplot bar chart with additional details

Great, we have plotted a bar chart in ggplot.

How it works...

In step 1, we import the pandas library and refer to it as pd. We also import the plotnine library. In step 2, we use read_csv to load the .csv file into a pandas dataframe, and then we call it houseprices_data.

In step 3, we inspect the dataset using the head method. to get a sense of the first five rows in the dataset; we also use the dtypes and shape attributes to get the data types and the number of rows and columns, respectively.

In step 4, we create the price per square meter variable. In step 5, we sort the dataframe based on the Price column using the sort_values method. This will be useful since we plan to visualize the top 10 areas based on house prices. In step 6, we plot a bar chart using ggplot. Within GGPLOT, first, we provide our data, then we provide the aesthetics, which in this case is our axes. Next, we provide the geom_bar chart type, which is for bar charts. For geom_bar, the height of the bars is determined by the stat parameter, and this is a count of the values on the *x* axis by default. We override the stat default and replace it with identity. This allows us to map the height of the bars to the raw values of our y variable (Price) instead of using a count. Also, we use the scale_x_discrete parameter to plot the bar chart in descending order based on price, just as it is within our sorted data. Lastly, we use theme to indicate our figure size.

In step 7, we provide additional details to make our bar chart informative. Using `labs`, we can provide title and axes labels, while with `theme`, we can provide the font size for the labels.

`ggplot` has a faceting capability that makes it possible to create multiple subset plots based on a specific variable in our data. However, subplots aren't directly supported in `GGPLOT` currently.

There's more...

`ggplot` supports several other geometric objects such as `geom_boxplot`, `geom_violin`, `geom_point`, `geom_histograms`, and so on. It also has aesthetically appealing themes, which can be very useful if the aesthetics of our charts is a key requirement.

See also

Here is an insightful article by *GeeksforGeeks* on the `plotnine` library:

`https://www.geeksforgeeks.org/data-visualization-using-plotnine-and-ggplot2-in-python/`

Visualizing data in Bokeh

Bokeh is another popular data visualization library in Python. Bokeh provides interactive and aesthetically appealing charts. These charts allow users to probe many scenarios interactively. Bokeh also supports custom JavaScript for special and advanced visualization use cases. With it, charts can be easily embedded into web pages.

Glyphs are the building blocks of bokeh visualizations. A glyph is a geometrical shape or marker that is used to represent data and create plots in bokeh. Typically, a plot consists of one or many geometrical shapes such as line, square, circle, rectangle, and so on. These shapes (glyphs) have visual information about the corresponding set of data.

Many of the important terms considered under the `matplotlib` recipe also apply to `bokeh`, and the syntax is also quite similar. We will explore `bokeh` through some examples.

Getting ready

We will continue working with Amsterdam House Prices Data from Kaggle.

We will also use the `bokeh` library for this recipe. You can install `bokeh` with `pip` using the following command:

```
pip install bokeh
```

How to do it...

We will learn how to visualize data using bokeh:

1. Import the pandas and bokeh libraries:

```
import pandas as pd
from bokeh.plotting import figure, show
import bokeh.plotting as bk_plot
from bokeh.io import output_notebook
output_notebook()
```

2. Load the .csv into a dataframe using read_csv. Then subset the dataframe to include only relevant columns:

```
houseprices_data = pd.read_csv("data/HousingPricesData.csv")

houseprices_data = houseprices_data[['Zip', 'Price', 'Area',
'Room']]
```

3. Inspect the data. Get a glimpse of the first five rows using the head method. Also get a sense of the data types as well as the number of columns and rows:

```
houseprices_data.head()
     Zip          Price    Area    Room
0    1091 CR      685000.0    64    3
1    1059 EL      475000.0    60    3
2    1097 SM      850000.0    109    4
3    1060 TH      580000.0    128    6
4    1036 KN      720000.0    138    5

houseprices_data.shape
(924,5)

houseprices_data.dtypes
Zip      object
Price      float64
Area      int64
Room      int64
```

4. Create a price per sqm variable based on the `Price` and `Area` variables:

```
houseprices_data['PriceperSqm'] = houseprices_data['Price']/
houseprices_data['Area']

houseprices_data.head()
     Zip           Price       Area    Room    PriceperSqm
0    1091 CR       685000.0    64      3       10703.125
1    1059 EL       475000.0    60      3       7916.6667
2    1097 SM       850000.0    109     4       7798.1651
3    1060 TH       580000.0    128     6       4531.25
4    1036 KN       720000.0    138     5       5217.3913
```

5. Sort the dataframe based on the house prices and inspect the output:

```
houseprices_sorted = houseprices_data.sort_values('Price',
ascending = False)
houseprices_sorted.head()
       Zip          Price        Area   Room     PriceperSqm
195    1017 EL      5950000.0    394    10       15101.5228
837    1075 AH      5850000.0    480    14       12187.5
305    1016 AE      4900000.0    623    13       7865.1685
103    1017 ZP      4550000.0    497    13       9154.9295
179    1012 JS      4495000.0    178    5        25252.8089
```

6. Plot a bar chart in bokeh with basic details:

```
data = houseprices_sorted[0:10]

fig = figure(x_range = data['Zip'],plot_width = 700, plot_height
= 500)
fig.vbar(x= data['Zip'], top = data['Price'], width = 0.9)
show(fig)
```

This results in the following output:

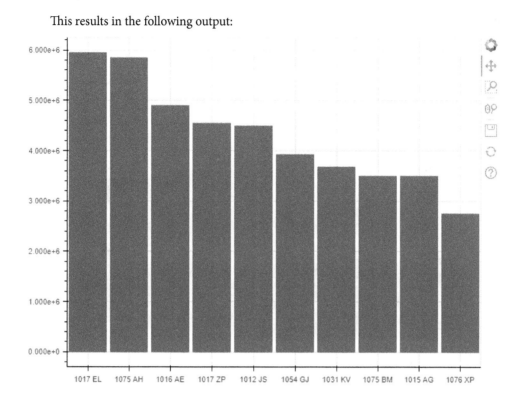

Figure 3.9: A bokeh bar chart with basic details

7. Plot a bar chart in bokeh with additional informative details:

```
fig = figure(x_range = data['Zip'],plot_width = 700, plot_height
= 500,
            title = 'Top 10 Areas with the highest house
prices', x_axis_label = 'Zip code',
            y_axis_label = 'House prices in millions')

fig.vbar(x= data['Zip'], top = data['Price'], width = 0.9)


fig.xaxis.axis_label_text_font_size = "15pt"
fig.xaxis.major_label_text_font_size = "10pt"
fig.yaxis.axis_label_text_font_size = "15pt"
fig.yaxis.major_label_text_font_size = "10pt"
fig.title.text_font_size = '15pt'
show(fig)
```

This results in the following output:

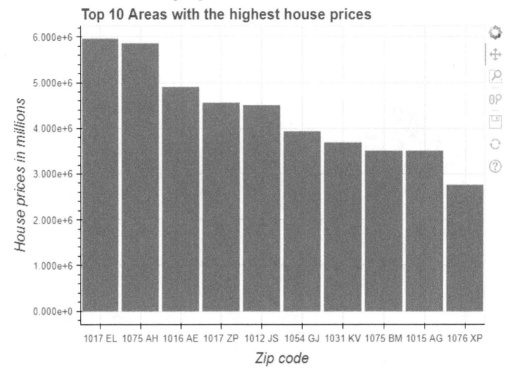

Figure 3.10: A bokeh bar chart with additional details

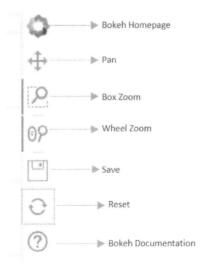

Figure 3.11: Default bokeh interactive features

8. Create subplots in bokeh to view multiple perspectives at once:

```
p1 = figure(x_range = data['Zip'],plot_width = 480, plot_height
= 400,
            title = 'Top 10 Areas with the highest house
prices', x_axis_label = 'Zip code',
            y_axis_label = 'House prices in millions')


p1.vbar(x= data['Zip'], top = data['Price'], width = 0.9)


p2 = figure(x_range = data['Zip'],plot_width = 480, plot_height
= 400,
            title = 'Top 10 Areas with the highest house prices
per sqm', x_axis_label = 'Zip code',
            y_axis_label = 'House prices per sqm')


p2.vbar(x= data['Zip'], top = data['PriceperSqm'], width = 0.9)


gp = bk_plot.gridplot(children=[[p1, p2]])
bk_plot.show(gp)
```

This results in the following output:

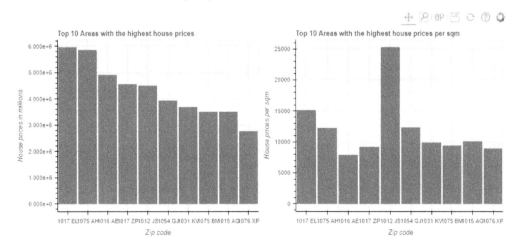

Figure 3.12: The bokeh subplots

Awesome! We have plotted a bar chart in bokeh.

How it works...

In step 1, we import the `pandas` library and refer to it as `pd`. We import the `plotting` module from the `bokeh` library and refer to it as `bk_plot`. We also import `figure`, `show`, and `output_notebook`. `figure` helps with defining our frame, `show` helps with displaying our charts, and `output_notebook` ensures our charts are displayed within our notebook. In step 2, we use `read_csv` to load the `.csv` file into a `pandas` dataframe. We call it `houseprices_data`.

In step 3, we inspect the dataset using the `head` method to get a sense of the first five rows in the dataset; we also use the `dtypes` and `shape` attributes to get the data types and the number of rows and columns respectively.

In step 4, we create the price per square meter variable. In step 5, we sort the dataframe based on the `Price` column using the `sort_values` method. This will be useful since we plan to visualize the top 10 areas based on house price. In step 6, we plot a bar chart using `bokeh`. We use the `figure` function to define the size of our plotting frame and also use the `x_range` parameter to specify the values of the *x* axis. Using the `vbar` function, we then plot our bar chart. The `x` parameter specifies the column for the *x* axis, while the `top` parameter specifies the height of the bars, and the `width` parameter specifies the size of the bar plot within the figure. In step 7, we provide additional details to make our bar chart informative. Within the figure, we specify our title and labels alongside the figure size. Next, we use the `title_text_font_size` function to provide a title. The `xaxis.axis_label_text_font_size` and `yaxis.axis_label_text_font_size` function provide the font size for our *x*- and *y*-axis labels, respectively. Then we use `xaxis.major_label_text_font_size` and `yaxis.major_label_text_font_size` to adjust the font size of the *x*- and *y*-axis values.

Bokeh provides interactive features such as zoom, pan, and hovering. In our chart, basic interactive features such as zoom and pan can be found on the top right. Pan allows us to move the plot or chart horizontally or vertically, and it is useful when exploring data with a wide range of values or when comparing different sections of a large dataset. Box zoom allows us to zoom in on a specific area of the plot or chart by dragging the cursor to draw a box around the area of interest. Wheel zoom allows us to zoom in and out of the plot or chart using the mouse wheel.

In step 8, we create subplots using the `gridplot` function. We assign the figures to the `p1` and `p2` variables. Next, we create our plots for each figure. Lastly, we use the `gridplot` function to create our subplots.

There's more...

Beyond pan, zoom, and hovering, Bokeh provides a wide range of interactive features. These include the following:

- **Interactions with callbacks**: Bokeh allows users to add custom JavaScript callbacks or Python callbacks to handle more complex user interactions

- **Interactivity with widgets**: Bokeh also provides widgets that can be used to create interactive dashboards

See also

Check out the `bokeh` user guide. It provides more information about `bokeh` capabilities and sample code:

- `https://docs.bokeh.org/en/latest/docs/user_guide.html`
- `https://docs.bokeh.org/en/3.0.2/docs/user_guide/interaction.html`

4

Performing Univariate Analysis in Python

When performing univariate analysis, we are usually interested in analyzing one or more variables in our dataset individually. Some insights we can glean during univariate analysis include the median, mode, maximum, range, and outliers. Univariate analysis can be performed on both categorical and numerical variables. Several chart options can be explored for both types of variables. These chart options can help us understand the underlying distribution of our data and identify any hidden patterns within the dataset. It is important to understand when to use each chart as this will ensure the accuracy of our analysis and the insights we derive from it.

In this chapter, we will cover the following:

- Performing univariate analysis using a histogram
- Performing univariate analysis using a boxplot
- Performing univariate analysis using a violin plot
- Performing univariate analysis using a summary table
- Performing univariate analysis using a bar chart
- Performing univariate analysis using a pie chart

Technical requirements

We will leverage the `pandas`, `matplotlib`, and `seaborn` libraries in Python for this chapter. The code and notebooks for this chapter are available on GitHub at `https://github.com/PacktPublishing/Exploratory-Data-Analysis-with-Python-Cookbook`.

Performing univariate analysis using a histogram

When visualizing one numeric variable in our dataset, there are various options to consider, and the histogram is one of them. A histogram is a bar graph-like representation that provides insights into our dataset's underlying frequency distribution, usually a continuous dataset. The x-axis of a histogram represents continuous values that have been split into bins or intervals while the y-axis represents the number or percentage of occurrences for each bin.

With the histogram, we can quickly identify outliers, data spread, skewness, and more.

In this recipe, we will explore how to create histograms in `seaborn`. The `histplot` method in `seaborn` can be used for this.

Getting ready

In this chapter, we will work with two datasets: the Amsterdam House Prices Data and the Palmer Archipelago (Antarctica) Penguins data, both from Kaggle.

Create a folder for this chapter and create a new Python script or Jupyter Notebook file in that folder. Create a data subfolder and place the `HousingPricesData.csv`, `penguins_size.csv`, and `penguins_lter.csv` files in that subfolder. Alternatively, you could retrieve all the files from the GitHub repository.

> **Note**
> Kaggle provides the Amsterdam House Prices and Penguins data for public use at `https://www.kaggle.com/datasets/thomasnibb/amsterdam-house-price-prediction` and `https://www.kaggle.com/datasets/parulpandey/palmer-archipelago-antarctica-penguin-data`. In this chapter, we will use the full datasets for the different recipes. The data is also available in the repository.

How to do it...

Let's learn how to create a histogram using the `seaborn` library:

1. Import the `pandas` and `seaborn` libraries:

    ```
    import pandas as pd
    import matplotlib.pyplot as plt
    import seaborn as sns
    ```

2. Load the .csv into a dataframe using read_csv. Subset the dataframe to include only the relevant columns:

```
penguins_data = pd.read_csv("data/penguins_size.csv")

penguins_data = penguins_data[['species','culmen_length_
mm']]
```

3. Check the first five rows using the head method. Check the number of columns and rows as well as the data types:

```
penguins_data.head()
        species    culmen_length_mm
0    Adelie     39.1
1    Adelie     39.5
2    Adelie     40.3
3    Adelie
4    Adelie     36.7

penguins_data.shape
(344, 4)

penguins_data.dtypes
species              object
culmen_length_mm     float64
```

4. Create a histogram using the histplot method:

```
sns.histplot( data = penguins_data, x= penguins_
data["culmen_length_mm"])
```

This results in the following output:

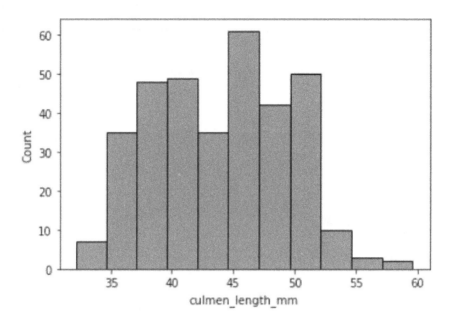

Figure 4.1: Seaborn histogram with basic details

5. Provide some additional details for the chart:

```
plt.figure(figsize= (12,6))
ax = sns.histplot( data = penguins_data, x= penguins_
data["culmen_length_mm"])
ax.set_xlabel('Culmen Length in mm',fontsize = 15)
ax.set_ylabel('Count of records', fontsize = 15)
ax.set_title('Univariate analysis of Culmen Length',
fontsize= 20)
```

This results in the following output:

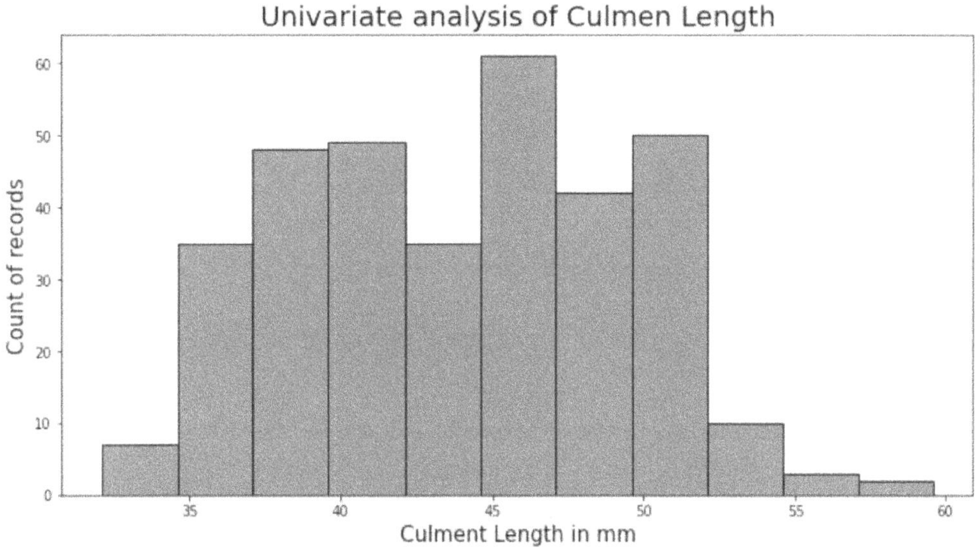

Figure 4.2: Seaborn histogram with additional details

Great. Now we have created a histogram on a single variable.

How it works...

In this recipe, we used the `pandas`, `matplotlib`, and `seaborn` libraries. In step 1, we import `pandas` and refer to it as `pd`, `matplotlib` as `plt` and `seaborn` as `sns`. In step 2, we use `read_csv` to load the `.csv` file into a `pandas` dataframe and call it `penguins_data`. We subset the dataframe to include only the two relevant columns. In step 3, we get a quick sense of what our data looks like by inspecting the first five rows using the `head` method. We also get a glimpse of the dataframe's shape (the number of rows and columns) and the data types using the `shape` and `dtypes` methods respectively.

In step 4, we create a histogram on the culmen length column using the `histplot` method in `seaborn`. In the method, we specify our dataset and the *x*-axis, which shows the culmen length. In step 5, we provide additional details for our chart, such as axis labels and titles.

Performing univariate analysis using a boxplot

Just like the histogram, the boxplot (also known as the whisker plot) is a good candidate for visualizing a single continuous variable within our dataset. Boxplots give us a sense of the underlying distribution of our dataset through five key metrics. The metrics include the minimum, first quartile, median, third quartile, and maximum values.

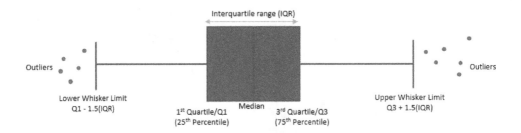

Figure 4.3: Boxplot illustration

In the preceding figure, we can see the following components of a boxplot:

- **The box:** This represents the interquartile range (25th percentile/1st quartile to the 75th percentile/3rd quartile). The median is the line within the box and it is also referred to as the 50th percentile.

- **The whisker limits:** The upper and lower whisker limits represent the range of values in our dataset which are not outliers. The position of the whiskers is calculated from the **interquartile range (IQR)**, 1st quartile, and 3rd quartile. This is represented by **Q1 - 1.5(IQR)** and **Q3 + 1.5(IQR)** for the lower and upper whisker respectively.

- **The circles:** The bottom and top circles outside of the whisker limits represent outliers in our dataset. These outliers are either unusually small values below the lower whisker limit, or unusually large values above the maximum whisker limit.

With boxplots, we can get insights into the spread of our dataset and easily identify outliers. In this recipe, we will explore how to create boxplots in seaborn. The boxplot method in seaborn can be used for this.

Getting ready

We will work with the Amsterdam House Prices Data from Kaggle in this recipe.

You can retrieve all the files from the GitHub repository.

How to do it...

Let's learn how to create a boxplot using the seaborn library:

1. Import the pandas and seaborn libraries:

    ```
    import pandas as pd
    import matplotlib.pyplot as plt
    import seaborn as sns
    ```

2. Load the `.csv` into a dataframe using `read_csv`. Then, subset the dataframe to include only the relevant columns:

```
houseprices_data = pd.read_csv("Data/HousingPricesData.
csv")


houseprices_data = houseprices_
data[['Zip','Price','Area','Room']]
```

3. Inspect the data. Check the first five rows using the `head` method. Also, check the data types as well as the number of columns and rows:

```
houseprices_data.head()
        Zip          Price      Area      Room
0    1091 CR    685000.0      64      3
1    1059 EL    475000.0      60      3
2    1097 SM    850000.0     109      4
3    1060 TH    580000.0     128      6
4    1036 KN    720000.0     138      5


houseprices_data.shape
(924, 4)


houseprices_data.dtypes
Zip         object
Price       float64
Area          int64
Room          int64
```

4. Create a boxplot using the `boxplot` method:

```
sns.boxplot(data = houseprices_data, x= houseprices_
data["Price"])
```

This results in the following output:

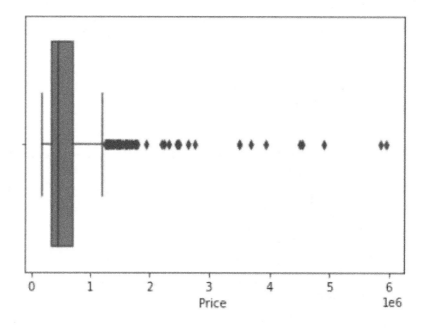

Figure 4.4: Seaborn boxplot with basic details

5. Provide some additional details for the chart:

```
plt.figure(figsize= (12,6))

ax = sns.boxplot(data = houseprices_data, x= houseprices_
data["Price"])
ax.set_xlabel('House Prices in millions',fontsize = 15)
ax.set_title('Univariate analysis of House Prices',
fontsize= 20)
plt.ticklabel_format(style='plain', axis='x')
```

This results in the following output:

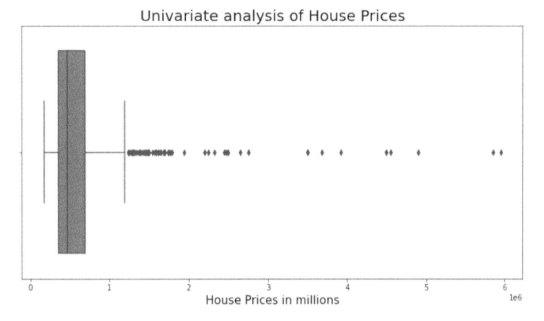

Figure 4.5: Seaborn boxplot with additional details

Awesome. We just created a boxplot on a single variable.

How it works...

In step 1, we import the pandas library and refer to it as pd. We also import the pyplot module from the matplotlib library. We refer to it as plt and import seaborn as sns. In step 2, we use read_csv to load the .csv file into a pandas dataframe. We call the dataframe houseprices_ data.

In step 3, we inspect the dataset using head to get a glimpse of the first 5 rows in the dataset; we also use the dtypes and shape attributes to get the data types, and the number of rows and columns respectively.

In step 4, we create a boxplot on house prices using the boxplot method in seaborn. In the method, we specify our dataset and the x-axis, which shows the house prices. Our boxplot reveals the spread of the price variable and outliers. In step 5, we provide additional details to our chart such as axis labels and titles. We also use the ticklabel_format method in matplotlib to change the scientific notation on the x-axis.

There's more...

Our boxplot seems to have several outliers. However, further analysis is typically required because univariate analysis doesn't always provide the full picture of what may be going on. That is why bivariate (analysis of 2 variables) and multivariate analyses (analysis of 3 or more variables) are also important. We may need to consider price along with size or price along with with location to get to get a sense of what is going on. We will cover this in more detail in the next chapter.

Performing univariate analysis using a violin plot

A violin plot is quite like a boxplot because it depicts the distribution of our dataset. The violin plot shows the peak of our data and where most values in our dataset are clustered. Just like boxplots provide summary statistics about our data, violin plots do the same along with providing additional information about the shape of our data.

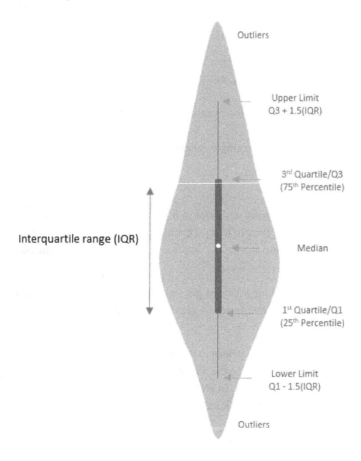

Figure 4.6: Violin plot illustration

In the preceding figure, we can see the following components of a violin plot:

- **The thick line:** This represents the interquartile range (25th percentile/1st quartile to the 75th percentile/3rd quartile).

- **The white dot:** This represents the median (50th percentile).

- **The thin line:** This is similar to the upper and lower whisker limits of the boxplot. It represents the range of values in our dataset that are not outliers. The lower and upper limits of the line are calculated from the IQR between the 1st quartile and 3rd quartile. This is represented by **Q1 - 1.5(IQR)** and **Q3 + 1.5(IQR)** for the lower and upper limits respectively.

- **The kernel density plot:** This displays the shape of the distribution of the underlying data. The data points within our dataset have a higher probability of being within the wider sections (peaks) and a lower probability of falling within the thinner sections (tails).

In this recipe, we will explore how to create violin plots in `seaborn`. The `violinplot` method in `seaborn` can be used for this.

Getting ready

We will work with the Amsterdam House Prices Data from Kaggle in this recipe.

You can retrieve all the files from the GitHub repository.

How to do it...

We will learn how to create a violin plot using the `seaborn` library:

1. Import the `pandas` and `seaborn` libraries:

    ```
    import pandas as pd
    import matplotlib.pyplot as plt
    import seaborn as sns
    ```

2. Load the `.csv` into a dataframe using `read_csv`. Then subset the dataframe to include only the relevant columns:

    ```
    houseprices_data = pd.read_csv("Data/HousingPricesData.
    csv")

    houseprices_data = houseprices_
    data[['Zip','Price','Area','Room']]
    ```

3. Get a glimpse of the first five rows using the head method. Also, check the data types as well as the number of columns and rows:

```
houseprices_data.head()
         Zip              Price      Area      Room
0     1091 CR        685000.0       64       3
1     1059 EL        475000.0       60       3
2     1097 SM        850000.0       109      4
3     1060 TH        580000.0       128      6
4     1036 KN        720000.0       138      5


houseprices_data.shape
(924, 4)


houseprices_data.dtypes
Zip           object
Price         float64
Area          int64
Room          int64
```

4. Create a violin plot using the violinplot method:

```
sns.violinplot(data = houseprices_data, x= houseprices_
data["Price"])
```

This results in the following output:

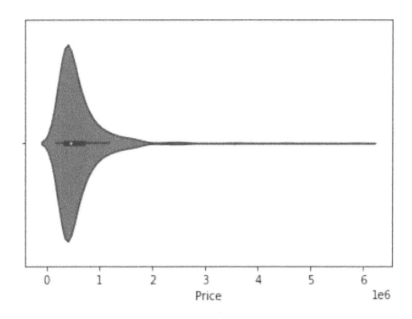

Figure 4.7: Seaborn violin plot with basic details

5. Provide some additional details for the chart:

```
plt.figure(figsize= (12,6))


ax = sns.violinplot(data = houseprices_data, x=
houseprices_data["Price"])
ax.set_xlabel('House Prices in millions',fontsize = 15)
ax.set_title('Univariate analysis of House Prices',
fontsize= 20)
plt.ticklabel_format(style='plain', axis='x')
```

This results in the following output:

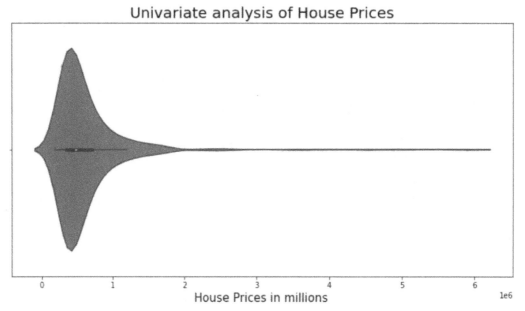

Figure 4.8: Seaborn violin plot with additional details

That's all. We have now created a violin plot on a single variable.

How it works...

In this recipe, we use the `pandas`, `matplotlib`, and `seaborn` libraries. In step 1, we import `pandas` and refer to it as pd. We also import the `matplotlib` library as plt and the `seaborn` library as sns. In step 2, we use `read_csv` to load the `.csv` file into a `pandas` dataframe and call it `houseprices_data`. We subset the dataframe to include only the four relevant columns. In step 3, we get a view of our data by inspecting the first five rows using the `head` method. We also get a glimpse of the dataframe shape (number of rows and columns) and the data types using the `shape` and `dtypes` methods respectively.

In step 4, we create a violin plot of our house price data using the `violinplot` method in `seaborn`. In the method, we specify our dataset and the *x*-axis, which shows the house prices. Our violin plot reveals the shape of the distribution of house prices within the data. The wider section is where most of the values in our dataset lie, while the thinner section indicates where fewer values lie.

In step 5, we provide additional details to our chart such as axis labels and title. We also use the `ticklabel_format` method in `matplotlib` to change the scientific notation on the *x*-axis.

Performing univariate analysis using a summary table

In univariate analysis, a summary table is very useful for analyzing numerical values within our dataset. In Python, these summary tables present statistics that summarize the central tendency, dispersion, and shape of the distribution of our dataset. The statistics covered include the count of non-empty records, mean, standard deviation, minimum, maximum, 25th percentile, 50th percentile and 75th percentile.

In pandas, the describe method provides these summary tables with all the aforementioned statistics.

Getting ready

We will work with the Amsterdam House Prices Data from Kaggle in this recipe.

You can retrieve all the files from the GitHub repository.

How to do it...

Let's learn how to create a summary table using the pandas library:

1. Import the pandas and seaborn libraries:

    ```
    import pandas as pd
    ```

2. Load the .csv into a dataframe using read_csv. Then subset the dataframe to include only the relevant columns:

    ```
    houseprices_data = pd.read_csv("Data/HousingPricesData.
    csv")

    houseprices_data = houseprices_data[['Zip','Price']]
    ```

3. Get a sense of the first five rows using the head method. Similarly, check out the dataframe's shape and the data types:

    ```
    houseprices_data.head()
                 Zip      Price
    0     1091 CR     685000.0
    1     1059 EL     475000.0
    2     1097 SM     850000.0
    3     1060 TH     580000.0
    4     1036 KN     720000.0
    ```

```
houseprices_data.shape
(924, 2)

houseprices_data.dtypes
Zip          object
Price        float64
```

4. Create a summary table using the describe method:

```
houseprices_data.describe()
      Price
count      920.0
mean     622065.42
std      538994.18
min      175000.0
25%      350000.0
50%      467000.0
75%      700000.0
max      5950000.0
```

This results in the following output:

	Price
count	9.200000e+02
mean	6.220654e+05
std	5.389942e+05
min	1.750000e+05
25%	3.500000e+05
50%	4.670000e+05
75%	7.000000e+05
max	5.950000e+06

Figure 4.9: pandas summary table

Awesome. We have now created a summary table on a single variable.

How it works...

In this recipe, we use the `pandas`, `matplotlib`, and `seaborn` libraries. In step 1, we import `pandas` and refer to it as `pd`, `matplotlib` as `plt`, and `seaborn` as `sns`. In step 2, we use `read_csv` to load the `.csv` file into a `pandas` dataframe and call it `houseprices_data`. We subset the dataframe to include only the four relevant columns. In step 3, we get a quick sense of what our data looks like by inspecting the first five rows using the `head` method. We also get a glimpse of the dataframe shape (number of rows and columns) and the data types using the `shape` and `dtypes` methods respectively.

In step 4, we create a summary table of our house price data using the `describe` method in `pandas`.

There's more...

The `describe` method can be used across multiple numerical values at once. This means once we are done selecting relevant columns, we can apply the `describe` method to view the summary statistics for each numerical column.

Performing univariate analysis using a bar chart

Like histograms, bar charts consist of rectangular bars. However, while the histogram analyzes numerical data, the bar chart analyzes categories. The *x*-axis typically represents the categories in our dataset while the *y*-axis represents the count of the categories or their occurrences by percentage. In some cases, the *y*-axis can also be the sum or average of a numerical column within our dataset. The bar chart provides quick insights especially when we need to quickly compare categories within our dataset.

In this recipe, we will explore how to create bar charts in `seaborn`. The `countplot` method in `seaborn` can be used for this. `seaborn` also has a `barplot` method. While `countplot` plots the count of each category, `barplot` plots a numeric variable against each category. This makes `countplot` better suited for univariate analysis, while the `barplot` method is better suited for bivariate analysis.

Getting ready

We will continue to work with the Palmer Archipelago (Antarctica) Penguin data from Kaggle.

How to do it...

We will learn how to create a bar plot using the `seaborn` library:

1. Import the `pandas` and `seaborn` libraries:

    ```
    import pandas as pd
    import matplotlib.pyplot as plt
    import seaborn as sns
    ```

2. Load the `.csv` into a dataframe using `read_csv`. Then subset the dataframe to include only the relevant columns:

    ```
    penguins_data = pd.read_csv("data/penguins_size.csv")


    penguins_data = penguins_data[['species','culmen_length_
    mm']]
    ```

3. Inspect the data. Check the first five rows using the `head` method. Also check the data types as well as the number of columns and rows:

    ```
    penguins_data.head()
            species     culmen_length_mm
    0    Adelie     39.1
    1    Adelie     39.5
    2    Adelie     40.3
    3    Adelie
    4    Adelie     36.7


    penguins_data.shape
    (344, 4)


    penguins_data.dtypes
    species               object
    culmen_length_mm      float64
    ```

4. Create a bar plot using the `countplot` method:

```
sns.countplot(data = penguins_data, x= penguins_
data['species'])
```

This results in the following output:

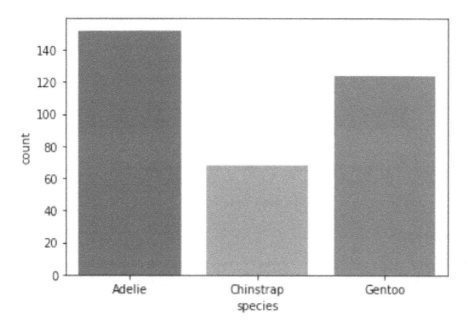

Figure 4.10: Seaborn bar chart with basic details

5. Provide some additional details for the chart:

```
plt.figure(figsize= (12,6))


ax = sns.countplot(data = penguins_data, x= penguins_
data['species'])
ax.set_xlabel('Penguin Species',fontsize = 15)
ax.set_ylabel('Count of records',fontsize = 15)
ax.set_title('Univariate analysis of Penguin Species',
fontsize= 20)ax.set_title('Univariate analysis of Culmen
Length', fontsize= 20)
```

This results in the following output:

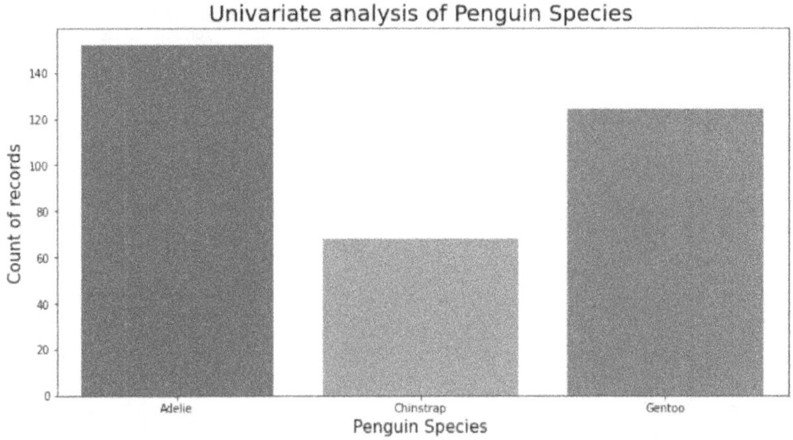

Figure 4.11: Seaborn bar chart with additional details

Great. We have now created a bar plot on a single variable.

How it works...

In step 1, we import the pandas library and refer to it as pd, along with the pyplot module from the matplotlib library, referring to it as plt, and seaborn as sns. In step 2, we use read_csv to load the .csv file into a pandas dataframe and call it penguins_data. We subset the dataframe to include only the two relevant columns. In step 3, we get a quick sense of what our data looks like by inspecting the first five rows using the head method. We also get a glimpse of the dataframe shape (number of rows and columns) and the data types using the shape and dtypes methods respectively.

In step 4, we create a bar chart comparing penguin species using the countplot method in seaborn. In the method, we specify our dataset and the x-axis which is the name of the species. In step 5, we provide additional details to our chart such as axis labels and title.

Performing univariate analysis using a pie chart

A pie chart is a circular visual that displays the relative sizes of various categories. Each slice of a pie chart represents a category and each category's size is proportional to its fraction of the total size of the data, typically 100 percent. Pie charts allow us to easily compare various categories.

In this recipe, we will explore how to create pie charts in matplotlib. seaborn doesn't have a pie chart method for creating pie charts, so, we will use matplotlib to achieve this. The pie method in matplotlib can be used for this.

Getting ready

We will continue to work with the Palmer Archipelago (Antarctica) Penguin data from Kaggle in this recipe.

How to do it...

Let's learn how to create a pie chart using the `matplotlib` library:

1. Import the `pandas` and `seaborn` libraries:

    ```
    import pandas as pd
    import matplotlib.pyplot as plt
    import seaborn as sns
    ```

2. Load the `.csv` into a dataframe using `read_csv`. Then subset the dataframe to include only the relevant columns:

    ```
    penguins_data = pd.read_csv("data/penguins_size.csv")


    penguins_data = penguins_data[['species','culmen_length_
    mm']]
    ```

3. Inspect the data. Check the first five rows using the `head` method. Also check the data types as well as the number of columns and rows:

    ```
    penguins_data.head()
             species    culmen_length_mm
    0    Adelie      39.1
    1    Adelie      39.5
    2    Adelie      40.3
    3    Adelie
    4    Adelie      36.7


    penguins_data.shape
    (344, 4)


    penguins_data.dtypes
    species                 object
    culmen_length_mm        float64
    ```

4. Group the data using the `groupby` method in `pandas`:

```
penguins_group = penguins_data.groupby('species').count()
penguins_group
species    culmen_length_mm
Adelie      151
Chinstrap     68
Gentoo      123
```

5. Reset the index using the `reset_index` method to ensure the index isn't the species column:

```
penguins_group= penguins_group.reset_index()
penguins_group
      species    culmen_length_mm
0     Adelie      151
1     Chinstrap     68
2     Gentoo      123
```

6. Create a pie chart using the `pie` method:

```
plt.pie(penguins_group["culmen_length_mm"], labels =
penguins_group['species'])
plt.show()
```

This results in the following output:

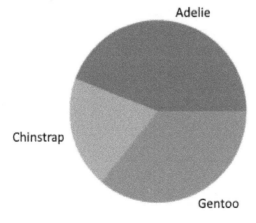

Figure 4.12: Matplotlib pie chart with basic details

7. Provide some additional details for the chart:

```
cols = ['g', 'b', 'r']
plt.pie(penguins_group["culmen_length_mm"], labels =
penguins_group['species'],colors = cols)
plt.title('Univariate Analysis of Species', fontsize=15)
plt.show()
```

This results in the following output:

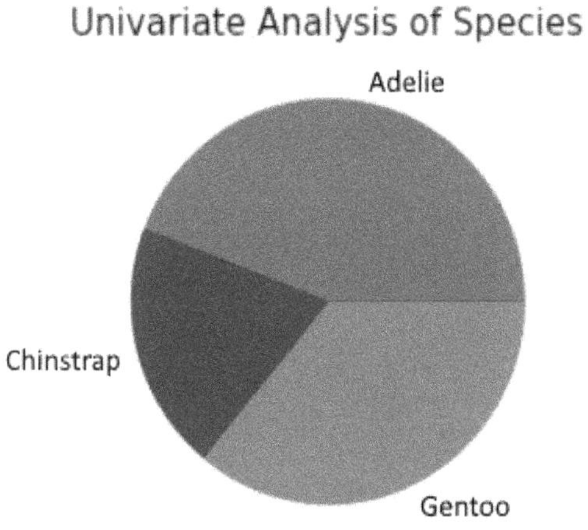

Figure 4.13: Matplotlib pie chart with additional details

Well done. We have now created a pie chart on a single variable.

How it works...

In step 1, we import pandas and refer to it as pd, along with matplotlib as plt and seaborn as sns. In step 2, we use read_csv to load the .csv file into a pandas dataframe and call it penguins_data. We subset the dataframe to include only the two relevant columns. In step 3, we get a quick sense of what our data looks like by inspecting the first five rows using the head method. We also get a glimpse of the dataframe shape (number of rows and columns) and the data types using the shape and dtypes methods respectively.

In step 4, we group our species column and calculate the count per species. In step 5, we use the reset_index method to reset the index of our output to ensure the index isn't the species column but the default index. In step 6, we create a pie chart using the pie method in matplotlib. The method takes the column containing penguin species counts as the first argument and species names as the second argument. In step 7, we provide additional details for our chart. We also specify the colors to be used via letters to signify the colors (g – green, b – blue, r – red).

5
Performing Bivariate Analysis in Python

Bivariate analysis uncovers insights embedded within two variables of interest. When performing this analysis, we are typically interested in how these two variables are distributed or related. Bivariate analysis can sometimes be more complex than univariate analysis because it involves the analysis of categorical and (or) numerical values. This means that in bivariate analysis, we can have three possible combinations of variables namely: numerical-numerical, numerical-categorical, and categorical-categorical.

It is important to understand the various chart options that cater to these combinations. These chart options can help us understand the underlying distribution of our data and identify any hidden patterns within the dataset.

In this chapter, we will cover the following key topics:

- Analyzing two variables using a scatter plot
- Creating crosstab/two-way tables on bivariate data
- Analyzing two variables using a pivot table
- Generating pairplots on two variables
- Analyzing two variables using a bar chart
- Generating boxplots for two variables
- Creating histograms on two variables
- Analyzing two variables using correlation analysis

Technical requirements

We will leverage the numpy, pandas, matplotlib, and seaborn libraries in Python for this chapter. The code and notebooks for this chapter are available on GitHub at https://github. com/PacktPublishing/Exploratory-Data-Analysis-with-Python-Cookbook.

Analyzing two variables using a scatter plot

A scatter plot clearly represents the relationship between two numerical variables. The numerical variables are plotted on the *x* and *y* axes, and the values plotted typically reveal a pattern. This pattern on a scatter plot can provide insights into the strength and direction of the relationship between the two variables. This could either be positive (i.e., as one variable increases, the other variable increases) or negative (i.e., as one variable increases, the other variable decreases). *Figure 5.1* demonstrates this further:

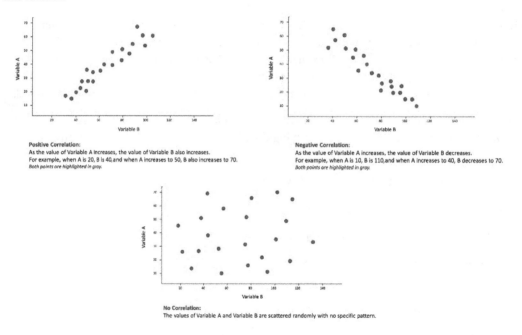

Figure 5.1: An illustration of a scatterplot

In this recipe, we will explore how to create scatter plots in seaborn. The scatterplot function in seaborn can be used for this.

Getting ready

We will work with only one dataset in this chapter: The Palmer Archipelago (Antarctica) penguin data from Kaggle.

Create a folder for this chapter and create a new Python script or Jupyter Notebook file in that folder. Create a data subfolder and place the `penguins_size.csv` and `penguins_lter.csv` files in that subfolder. Alternatively, you can retrieve all the files from the GitHub repository.

> **Note**
>
> Kaggle provides the Penguins data for public use at `https://www.kaggle.com/datasets/parulpandey/palmer-archipelago-antarctica-penguin-data`. In this chapter, we will use the full dataset for different recipes. The data is also available in the repository.
>
> **Citation**
>
> Gorman KB, Williams TD, Fraser WR (2014), *Ecological Sexual Dimorphism and Environmental Variability within a Community of Antarctic Penguins* (Genus Pygoscelis). *PLoS ONE* 9(3): e90081. doi:10.1371/journal.pone.0090081

How to do it...

We will learn how to create a scatter plot using the `seaborn` library:

1. Import the `pandas`, `matplotlib`, and `seaborn` libraries:

   ```
   import pandas as pd
   import matplotlib.pyplot as plt
   import seaborn as sns
   ```

2. Load the `.csv` file into a dataframe using `read_csv`. Subset the dataframe to include only the relevant columns:

   ```
   penguins_data = pd.read_csv("data/penguins_size.csv")

   penguins_data = penguins_data[['species','culmen_length_mm',
   'body_mass_g']]
   ```

3. Check the first five rows using the `head` method. Check the number of columns and rows as well as the data types:

   ```
   penguins_data.head()
          species   culmen_length_mm body_mass_g
   ```

```
0       Adelie      39.1                3750.0
1       Adelie      39.5                3800.0
2       Adelie      40.3                3250.0
3       Adelie
4       Adelie      36.7                3450.0


penguins_data.shape
(344, 3)


penguins_data.dtypes
species                 object
culmen_length_mm        float64
body_mass_g             float64
```

4. Create a scatter plot using the `scatterplot` method:

```
sns.scatterplot(data = penguins_data, x= penguins_data["culmen_
length_mm"], y= penguins_data['body_mass_g'])
```

This results in the following output:

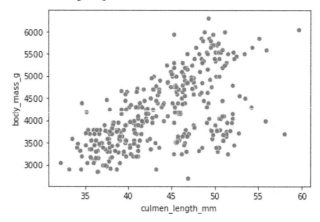

Figure 5.2: A scatterplot with basic details

5. Set the chart size and chart title:

```
plt.figure(figsize= (12,6))


ax = sns.scatterplot(data = penguins_data, x= penguins_
data["culmen_length_mm"], y= penguins_data['body_mass_g'])
ax.set_title('Bivariate analysis of Culmen Length and body
mass', fontsize= 20)
```

This results in the following output:

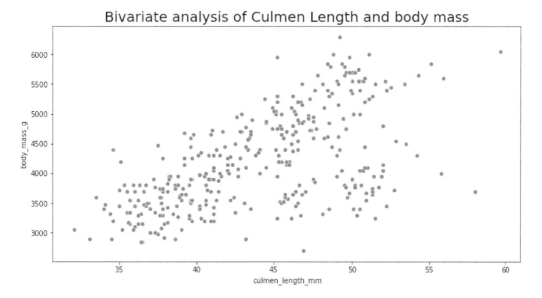

Figure 5.3: A scatterplot with additional details

Well done. We have created a scatter plot on two variables of interest.

How it works...

In this recipe, we use the pandas, matplotlib, and seaborn libraries. In *step 1*, we import pandas and refer to it as pd, matplotlib as plt, and seaborn as sns. In *step 2*, we use read_csv to load the .csv file into a pandas dataframe and call it penguins_data. We subset the dataframe to include only three relevant columns. In *step 3*, we get a quick sense of what our data looks like by inspecting the first five rows using the head method. We also get a glimpse of the dataframe shape (number of rows and columns) and the data types using the shape and dtypes attributes, respectively.

In *step 4*, we create a scatter plot on the two variables of interest using the scatterplot function in seaborn. In the function, we specify our dataset, the *x* axis, which is the culmen length, and the *y* axis, which is the body mass. In *step 5*, we provide additional details to our chart, such as the title.

There's more...

The scatterplot method has a hue parameter, which can take an additional variable. This variable typically must be a categorical variable. The hue parameter assigns a unique color to each category within the categorical variable. This means the points on the plot will take on the color of the category they belong to. With this option, analysis can be performed across three variables.

See also...

To get more information about the strength of the relationship between two numerical variables, check out the *Analysing two variables using correlation analysis* recipe in this chapter.

Creating a crosstab/two-way table on bivariate data

A crosstab displays the relationship between categorical variables in a matrix format. The rows of the crosstab are typically the categories within the first categorical variable, while the columns are the categories of the second categorical variable. The values within the crosstab are either the frequency of occurrence or the percentage of occurrence. It is also known as a two-way table or contingency table. With the crosstab, we can easily uncover trends and patterns, especially as they relate to specific categories within our dataset.

In this recipe, we will explore how to create crosstabs in pandas. The crosstab method in pandas can be used for this.

Getting ready

We will work with the Palmer Archipelago (Antarctica) penguin data from Kaggle in this recipe. You can retrieve all the files from the GitHub repository.

How to do it...

We will learn how to create a crosstab using the pandas library:

1. Import the pandas library:

    ```
    import pandas as pd
    ```

2. Load the .csv file into a dataframe using read_csv. Then, subset the dataframe to include only the relevant columns:

    ```
    penguins_data = pd.read_csv("data/penguins_size.csv")
    penguins_data = penguins_data[['species','culmen_length_
    mm','sex']]
    ```

3. Inspect the data. Check the first five rows using the head method. Also, check the data types as well as the number of columns and rows:

    ```
    penguins_data.head()
          species    culmen_length_mm    sex
    0     Adelie          39.1            MALE
    ```

```
1      Adelie        39.5                FEMALE
2      Adelie        40.3                FEMALE
3      Adelie
4      Adelie        36.7                FEMALE
penguins_data.shape
(344, 3)


penguins_data.dtypes
species               object
culmen_length_mm      float64
sex                   object
```

4. Create a crosstab using the `crosstab` function:

```
pd.crosstab(index= penguins_data['species'], columns= penguins_
data['sex'])
```

This results in the following output:

sex	FEMALE	MALE
species		
Adelie	73	73
Chinstrap	34	34
Gentoo	58	62

Figure 5.4: A pandas crosstab

Awesome. We have created a crosstab.

How it works...

In this recipe, we use the `pandas` library. In *step 1*, we import the `pandas` library and refer to it as pd. In step 2, we load the `.csv` file into a `pandas` dataframe using `read_csv`, and we call the `penguins_data` dataframe. We also subset it only for the relevant columns.

In step 3, we use the `head` method to get a sense of the first five rows in the dataset. Using the `dtypes` attribute, we get a view of the data types, and using the `shape` attribute, we get a view of the dataframe shape (number of rows and columns).

In step 4, we create a crosstab using the `crosstab` function in `pandas`. In the method, we use the `index` parameter to specify the rows of the crosstab and use the `columns` parameter to specify the columns of the crosstab.

Analyzing two variables using a pivot table

A pivot table summarizes our dataset by grouping and aggregating variables within the dataset. Some of the aggregation functions within the pivot table include a sum, count, average, minimum, maximum, and so on. For bivariate analysis, the pivot table can be used for categorical-numerical variables. The numerical variable is aggregated for each category in the categorical variable.

The name pivot table has its origin in spreadsheet software. The summary provided by a pivot table can easily uncover meaningful insights from a large dataset.

In this recipe, we will explore how to create a pivot table in `pandas`. The `pivot_table` function in `pandas` can be used for this.

Getting ready

We will work with the Palmer Archipelago (Antarctica) penguin data from Kaggle in this recipe. You can retrieve all the files from the GitHub repository.

How to do it...

We will learn how to create a pivot table using the `pandas` library:

1. Import the numpy and pandas libraries:

    ```
    import numpy as np
    import pandas as pd
    ```

2. Load the `.csv` file into a dataframe using `read_csv`. Then, subset the dataframe to include only the relevant columns:

    ```
    penguins_data = pd.read_csv("data/penguins_size.csv")
    penguins_data = penguins_data[['species','culmen_length_
    mm','sex']]
    ```

3. Inspect the data. Check the first five rows using the `head` method. Also, check the data types as well as the number of columns and rows:

    ```
    penguins_data.head()
    ```

    ```
    penguins_data.shape
    ```

    ```
    penguins_data.dtypes
    ```

4. Create a pivot table using the `pivot_table` function:

```
pd.pivot_table(penguins_data, values='culmen_length_mm',
                        index='species',
                        aggfunc=np.mean)
```

This results in the following output:

culmen_length_mm	
species	
Adelie	38.791391
Chinstrap	48.833824
Gentoo	47.504878

Figure 5.5: A pandas pivot table

That's all. We have now created a pivot table.

How it works...

In step 1, we import the `pandas` library and refer to it as `pd`. We also import the `numpy` library and refer to it as `np`. In step 2, we use `read_csv` to load the `.csv` file into a `pandas` dataframe. We call the `penguins_data` dataframe and subset it only for the relevant columns.

In step 3, we get a glimpse of the first five rows of our dataset by using the `head` method. Using the `dtypes` attribute, we get a view of the data types, and using the `shape` method, we get a view of the dataframe shape (number of rows and columns).

In step 4, we create a pivot table using the `pivot_table` function in `pandas`. In the function, we use the `index` parameter to specify the rows of the pivot table and use the `values` parameter to specify the numeric column values of the pivot table. The `aggfunc` parameter specifies the aggregation function to be applied to the numeric column values. We use the `numpy` `mean` function as the aggregation function.

There is more...

The `pivot_table` function has a `columns` Parameter, which can take an additional variable. Typically, this variable must be a categorical variable. The categories of this categorical variable are added as columns to the pivot table just like a crosstab. The values within each column will remain the values of our aggregated numerical variable. A simple example of this is having the categories of the `species` variable as the rows, having the categories of the `sex` variable as columns, and aggregating the `culmen_length` variable to generate the values of the pivot table. This is highlighted in the following figure:

sex	FEMALE	MALE
species		
Adelie	37.257534	40.390411
Chinstrap	46.573529	51.094118
Gentoo	45.563793	49.393548

Figure 5.6: A pandas pivot table with two categorical variables

Generating pairplots on two variables

The pairplot provides a visual representation of the distribution of single variables and the relationship between two variables. The pairplot isn't just one plot but a group of subplots that display single variables (numerical or categorical) and two variables (numerical-numerical and categorical-numerical). A pairplot combines charts such as histograms, density plots, and scatter plots to represent the variables in the dataset. It provides a simple way to view the distribution and relationship across multiple variables in our dataset.

In this recipe, we will explore how to create pairplots in `seaborn`. The `pairplot` function in `seaborn` can be used for this.

Getting ready

We will work with the Palmer Archipelago (Antarctica) penguin data from Kaggle in this recipe. You can retrieve all the files from the GitHub repository.

How to do it...

We will learn how to create a pairplot using the `seaborn` library:

1. Import the `pandas`, `matplotlib`, and `seaborn` libraries:

    ```
    import pandas as pd
    import matplotlib.pyplot as plt
    import seaborn as sns
    ```

2. Load the .csv into a dataframe using read_csv. Then, subset the dataframe to include only the relevant columns:

```
penguins_data = pd.read_csv("data/penguins_size.csv")
penguins_data = penguins_data[['species','culmen_length_
mm','body_mass_g','sex']]
```

3. Get a sense of the first five rows using the head method. Also, get a sense of the dataframe shape as well as the data types:

```
penguins_data.head()
   species   culmen_length_mm   body_mass_g   sex
0   Adelie       39.1              3750.0      MALE
1   Adelie       39.5              3800.0      FEMALE
2   Adelie       40.3              3250.0      FEMALE
3   Adelie
4   Adelie       36.7              3450.0      FEMALE

penguins_data.shape
(344, 4)

penguins_data.dtypes
species                 object
culmen_length_mm        float64
body_mass_g             float64
sex                     object
```

4. Create a pairplot using the pairplot function:

```
sns.pairplot( data = penguins_data)
```

This results in the following output:

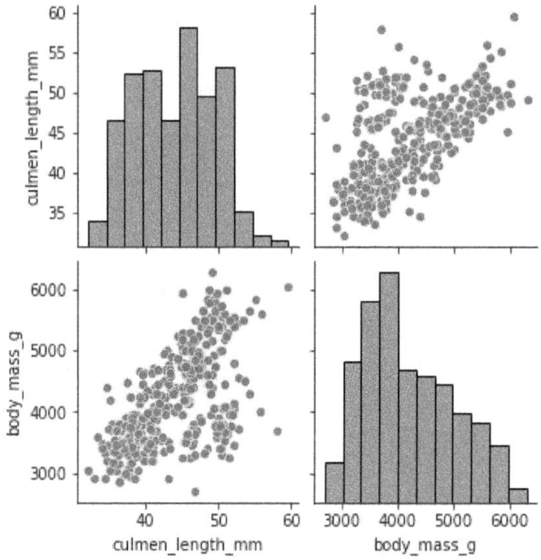

Figure 5.7: A pairplot with univariate and bivariate analysis

Great. We have now created a pairplot on our dataset.

How it works...

For the pairplot, we use the `pandas`, `matplotlib`, and `seaborn` libraries. In step 1, we import the three libraries. We import `pandas` and refer to it as `pd`, and then we import `pyplot` in `matplotlib` and refer to it as `plt`, and `seaborn` as `sns`. In step 2, we use `read_csv` to load the `.csv` file into a `pandas` dataframe and call it `penguins_data`. We subset the dataframe to include only four relevant columns.

In step 3, we get a quick sense of what our data looks like by inspecting the first five rows using the `head` method. We also get a glimpse of the dataframe shape (number of rows and columns) and the data types using the `shape` and `dtypes` attributes, respectively.

In step 4, we create a pair plot using the `pairplot` function in `pandas`. In the function, we use the `data` parameter to specify the data to be used for the pairplot.

Analyzing two variables using a bar chart

The bar chart can be used for both univariate and bivariate analysis. For bivariate analysis, the *x* axis typically represents the categories in our dataset, while the *y* axis represents a numerical variable. This means the bar chart is usually used for categorical-numerical analysis. The numerical variable is typically aggregated using functions such as the sum, median, mean, and so on. A bar chart provides quick insights, especially when we need to quickly compare categories within our dataset.

In this recipe, we will explore how to create bar charts for bivariate analysis in `seaborn`. `seaborn` has a `barplot` function that is used for this. `seaborn` also has a `countplot` function that plots a bar chart; however, the `countplot` function only plots the count of each category. Therefore, it is used for univariate analysis. The `barplot` function, on the other hand, plots a numerical variable against each category.

Getting ready

We will continue working with the Palmer Archipelago (Antarctica) penguin data from Kaggle in this recipe. You can retrieve all the files from the GitHub repository.

How to do it...

We will learn how to create a bar chart using the `seaborn` library:

1. Import the `numpy`, `pandas`, `matplotlib`, and `seaborn` libraries:

    ```
    import numpy as np
    import pandas as pd
    import matplotlib.pyplot as plt
    import seaborn as sns
    ```

2. Load the `.csv` into a dataframe using `read_csv`. Then, subset the dataframe to include only the relevant columns:

    ```
    penguins_data = pd.read_csv("data/penguins_size.csv")
    penguins_data = penguins_data[['species','culmen_length_
    mm','sex']]
    ```

3. Inspect the data. Check the first five rows using the `head` method. Also, check the data types as well as the number of columns and rows:

    ```
    penguins_data.head()

    penguins_data.shape

    penguins_data.dtypes
    ```

4. Create a bar chart using the `barplot` function:

```
sns.barplot(data = penguins_data,x=penguins_
data['species'],y=penguins_data['culmen_length_mm'],estimator =
np.median)
```

This results in the following output:

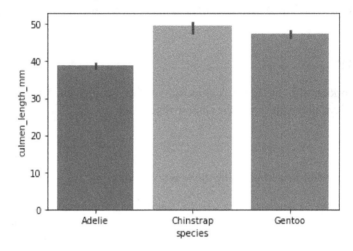

Figure 5.8: A bar chart with basic details

5. Set the chart size and chart title. Also, specify the axes' label names and label sizes:

```
plt.figure(figsize= (12,6))

ax = sns.barplot(data = penguins_data,x=penguins_
data['species'],y=penguins_data['culmen_length_mm'],estimator =
np.median)
ax.set_xlabel('Species',fontsize = 15)
ax.set_ylabel('Culmen_length_mm', fontsize = 15)
ax.set_title('Bivariate analysis of Culmen Length and Species',
fontsize= 20)
```

This results in the following output:

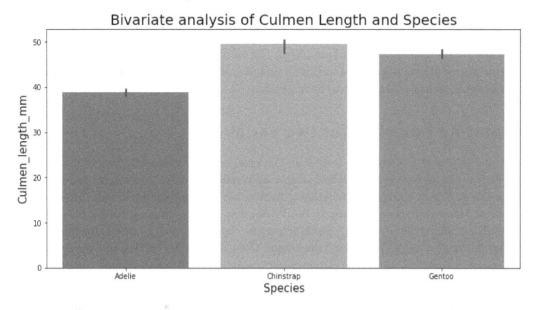

Figure 5.9: Bar chart with additional details

Well done. We have now created a bar chart based on two variables of interest.

How it works...

In step 1, we import the numpy and pandas libraries and refer to them as np and pd, respectively. We also import the pyplot module from the matplotlib library and import seaborn. We refer to pyplot as plt and seaborn as sns. In step 2, we use read_csv to load the .csv file into a pandas dataframe. We call the penguins_data dataframe and subset it for only relevant columns.

In step 3, we inspect the dataset using head to get a sense of the first five rows in the dataset. Using the dtypes attribute, we get a view of the data types, and using the shape attribute, we get a view of the dataframe shape (number of rows and columns).

In step 4, we create a bar chart using the barplot function in seaborn. In the function, we specify our dataset and the *x* and *y* axes. The *x* axis is the species variable, while the *y* axis is the culmen length. We also specify the numpy median function as the aggregation function for the *y* axis using the estimator parameter.

There is more...

The `barplot` function has a hue parameter, which can take an additional variable. Typically, this variable must be a categorical variable. With the hue parameter, multiple bars are added per category to the *x* axis. The number of bars added per category on the *x* axis is usually based on the number of categories in our additional categorical variable. A unique color is also assigned to each bar. With this option, analysis can be performed across three variables.

Generating box plots for two variables

A boxplot can be used for univariate analysis and bivariate analysis. When analyzing two variables, a boxplot is useful for analyzing numerical-categorical variables. Just like in univariate analysis, the boxplot also gives us a sense of the underlying distribution of a continuous variable through five key metrics. However, in bivariate analysis, the distribution of the continuous variable is displayed across each category of the categorical variable of interest. The five key metrics include the minimum, first quartile, median, third quartile, and maximum. These metrics give insights into the spread of our dataset and possible outliers. The boxplot is explained in more detail in *Chapter 4, Performing Univariate Analysis in Python*.

In this recipe, we will explore how to create boxplots in `seaborn`. The `boxplot` function in `seaborn` can be used for this.

Getting ready

We will continue working with the Palmer Archipelago (Antarctica) penguin data from Kaggle in this recipe. You can retrieve all the files from the GitHub repository.

How to do it...

We will learn how to create a boxplot using the `seaborn` library:

1. Import the `pandas`, `matplotlib`, and `seaborn` libraries:

    ```
    import pandas as pd
    import matplotlib.pyplot as plt
    import seaborn as sns
    ```

2. Load the `.csv` into a dataframe using `read_csv`. Then, subset the dataframe to include only the relevant columns:

    ```
    penguins_data = pd.read_csv("data/penguins_size.csv")
    penguins_data = penguins_data[['species','culmen_length_
    mm','sex']]
    ```

3. Inspect the data. Check the first five rows using the `head` method. Also, check the data types as well as the number of columns and rows:

```
penguins_data.head()

penguins_data.shape

penguins_data.dtypes
```

4. Create a boxplot using the `boxplot` function:

```
sns.boxplot( data = penguins_data, x= penguins_data['species']
,  y= penguins_data["culmen_length_mm"])
```

This results in the following output:

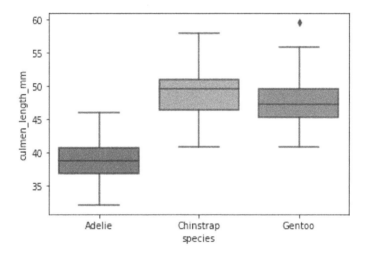

Figure 5.10: A box plot with basic details

5. Set the chart size and chart title. Also, specify the axes' label names and label sizes:

```
plt.figure(figsize= (12,6))

ax = sns.boxplot( data = penguins_data, x= penguins_
data['species']  ,   y= penguins_data["culmen_length_mm"])
ax.set_xlabel('Culmen Length in mm',fontsize = 15)
ax.set_ylabel('Count of records', fontsize = 15)
ax.set_title('Bivariate analysis of Culmen Length and Species',
fontsize= 20)
```

This results in the following output:

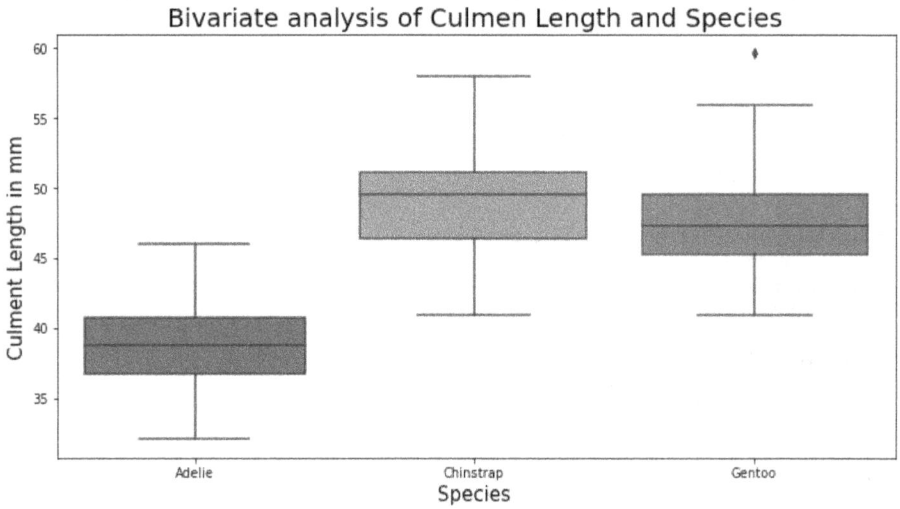

Figure 5.11: Box plot with additional details

That's all. We have now created a boxplot based on two variables.

How it works...

For the boxplot, we use the `pandas`, `matplotlib`, and `seaborn` libraries. In step 1, we import `pandas` and refer to it as `pd`, and then we import `pyplot` from `matplotlib` and refer to it as `plt`, and `seaborn` as `sns`. In step 2, we load the `.csv` file into a `pandas` dataframe using the `read_csv` function and call it `penguins_data`. We then subset the dataframe to include only the two relevant columns.

In step 3, we use `head` to get a view of the first five rows of our dataset. We also get a glimpse of the dataframe shape (number of rows and columns) and the data types using the `shape` and `dtypes` attributes, respectively.

In step 4, we create a boxplot using the `boxplot` method in `seaborn`. In the method, we specify our dataset and the axes. The *x* axis is the *species* variable, while the *y* axis is the culmen length variable. In step 5, we provide additional details to our chart such as the axes' labels and the title.

Creating histograms on two variables

Just like a boxplot, a histogram can also be used for univariate analysis and bivariate analysis. For bivariate analysis, a histogram is useful for analyzing numerical-categorical variables. It is usually straightforward when our categorical variable has only two categories. However, it becomes complex when we have more than two categories. Typically, in bivariate analysis using histograms, we can overlay the histogram for each of the categories over each other and assign a specific color to the histogram representing each category. This helps us to easily identify distinct distributions of our continuous variable across categories in our categorical variable of interest. This approach is best only for categorical variables with at most three categories.

In this recipe, we will explore how to create histograms in `seaborn`. The `histplot` method in `seaborn` can be used for this.

Getting ready

We will continue working with the Palmer Archipelago (Antarctica) penguin data from Kaggle in this recipe. You can retrieve all the files from the GitHub repository.

How to do it...

We will learn how to create a histogram using the `seaborn` library:

1. Import the `pandas`, `matplotlib`, and `seaborn` libraries:

    ```
    import pandas as pd
    import matplotlib.pyplot as plt
    import seaborn as sns
    ```

2. Load the `.csv` into a dataframe using `read_csv`. Then, subset the dataframe to include only the relevant columns:

    ```
    penguins_data = pd.read_csv("data/penguins_size.csv")
    penguins_data = penguins_data[['species','culmen_length_
    mm','sex']]
    ```

3. Inspect the data. Check the first five rows using the `head` method. Also, check the data types as well as the number of columns and rows:

    ```
    penguins_data.head()


    penguins_data.shape


    penguins_data.dtypes
    ```

4. Create a histogram using the `histplot` method:

    ```
    penguins_data_male = penguins_data.loc[penguins_data['sex']==
    'MALE',:]
    penguins_data_female = penguins_data.loc[penguins_data['sex']==
    'FEMALE',:]
    ```

```
sns.histplot( data = penguins_data_male, x= penguins_data_
male["culmen_length_mm"], alpha=0.5, color = 'blue')
sns.histplot( data = penguins_data_female, x= penguins_data_
female["culmen_length_mm"], alpha=0.5, color = 'red')
```

This results in the following output:

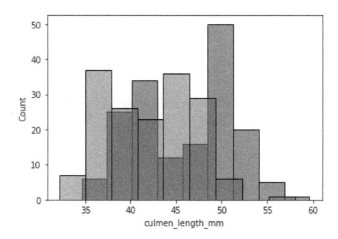

Figure 5.12: A histogram with basic details

5.   Set the chart size and chart title. Also specify the axes' label names and label sizes:

```
plt.figure(figsize= (12,6))

ax = sns.histplot( data = penguins_data_male, x= penguins_data_
male["culmen_length_mm"], alpha=0.5, color = 'blue')
ax = sns.histplot( data = penguins_data_female, x= penguins_
data_female["culmen_length_mm"], alpha=0.5, color = 'red')

ax.set_xlabel('Culmen Length in mm',fontsize = 15)
ax.set_ylabel('Count of records', fontsize = 15)
ax.set_title('Bivariate analysis of Culmen Length and Sex',
fontsize= 20)
plt.legend(['Male Penguins', 'Female Penguins'],loc="upper
right")
```

This results in the following output:

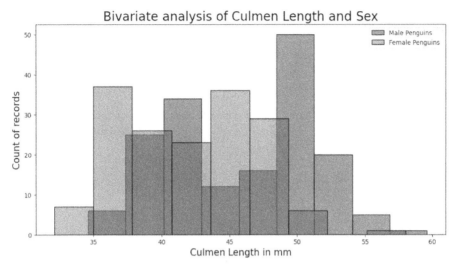

Figure 5.13: Histogram with additional details

Awesome. We just created a histogram based on two variables of interest.

How it works...

In step 1, we use the pandas, matplotlib, and seaborn libraries. We import pandas and refer to it as pd, and then we import pyplot from matplotlib as plt, and seaborn as sns. In step 2, we use read_csv to load the .csv file into a pandas dataframe and call it penguins_data. We subset the dataframe to include only the three relevant columns.

In step 3, we get a view of our data by inspecting the first five rows using the head method. We also get a glimpse of the dataframe shape (number of rows and columns) and the data types using the shape and dtypes methods, respectively.

In step 4, we create a histogram using the histplot method in seaborn. We start by creating two datasets, one for the male and the other for the female species. We then create two separate histograms for each dataset. In each histplot method, we specify the dataset and the *x* axis, which is the culmen length. We use the alpha argument to define the transparency of each histogram; a transparency of less than 0.6 works well when creating multiple charts, as shown previously. Lastly, we use the color argument to define the color of each histogram.

In step 5, we provide additional details to our chart, such as the axes' labels and the title.

Analyzing two variables using a correlation analysis

Correlation measures the strength of the relationship between two numerical variables. The strength is usually within a range of -1 and +1. -1 is a perfect negative linear correlation, while +1 is a perfect positive linear correlation, and 0 means no correlation. Correlation uncovers hidden patterns and insights in our dataset. In the case of a positive correlation, when one variable increases, the other also increases, and in the case of a negative correlation, when one variable increases, the other decreases.

In addition, a high correlation (a high correlation value) means there is a strong relationship between the two variables, while a low correlation means there is a weak relationship between the variables. It is important to note that correlation doesn't mean causation.

In this recipe, we will explore how to create correlation heatmaps in `seaborn`. The `heatmap` method in `seaborn` can be used for this.

Getting ready

We will continue working with the Palmer Archipelago (Antarctica) penguin data from Kaggle in this recipe. You can retrieve all the files from the GitHub repository.

How to do it...

We will learn how to create a correlation heatmap using the `seaborn` library:

1. Import the `pandas`, `matplotlib`, and `seaborn` libraries:

    ```
    import pandas as pd
    import matplotlib.pyplot as plt
    import seaborn as sns
    ```

2. Load the `.csv` into a dataframe using `read_csv`. Then, subset the dataframe to include only the relevant columns:

    ```
    penguins_data = pd.read_csv("data/penguins_size.csv")

    penguins_data = penguins_data[['species','culmen_length_mm',
    'body_mass_g']]
    ```

3. Inspect the data. Check the first five rows using the `head` method. Also, check the data types as well as the number of columns and rows:

    ```
    penguins_data.head()
    penguins_data.shape

    penguins_data.dtypes
    ```

4. Create a correlation heatmap chart using the `heatmap` method:

```
penguins_corr = penguins_data.corr()
heatmap = sns.heatmap(penguins_corr, vmin=-1, vmax=1,
annot=True)
```

This results in the following output:

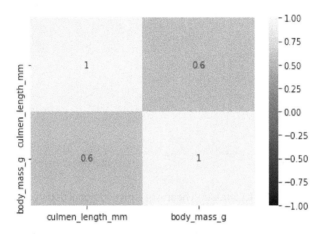

Figure 5.14: A correlation heatmap with basic details

5. Set the chart size and chart title. Also, set a padding value to define the distance of the title from the top of the heatmap:

```
plt.figure(figsize=(16, 6))
heatmap = sns.heatmap(penguins_data.corr(), vmin=-1, vmax=1,
annot=True)
heatmap.set_title('Correlation Heatmap',
fontdict={'fontsize':12}, pad=12);
```

This results in the following output:

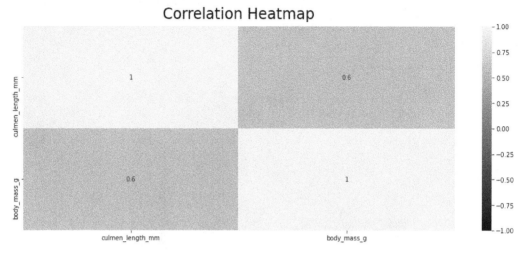

Figure 5.15: A correlation heatmap with additional details

Well done. Now we have created a correlation heatmap on our dataset.

How it works...

In step 1, we import the `pandas` library and refer to it as `pd`, and we import `pyplot` from `matplotlib` as `plt`, and `seaborn` as `sns`. In step 2, load the `.csv` file into a `pandas` dataframe using the `read_csv` method, and we call the `penguins_data` dataframe. We then subset it for only relevant columns.

In step 3, we use the `head` method to get a quick sense of what the first five rows look like. Using the `dtypes` method, we get a view of the data types, and using the `shape` method, we get a view of the dataframe shape (number of rows and columns).

In step 4, we create a heatmap using the `heatmap` method. Using the `corr` method in `pandas`, we define a correlation matrix and assign it to a variable. We then specify the correlation matrix as the dataset within the `heatmap` method. Using the `vmin` and `vmax` arguments, we specify the minimum and maximum values to be displayed on the heatmap. Using the `annot` argument, we set the annotation to `True` to display the correlation values on the heatmap. In step 5, we add more details to the chart. We use the `set_title` method to set a title. We use the `fontdict` argument to set the font size and use the `pad` argument to define the distance of the title from the top of the heatmap.

6

Performing Multivariate Analysis in Python

A common problem we will typically face with large datasets is analyzing multiple variables at once. While techniques covered under univariate and bivariate analysis are useful, they typically fall short when we are required to analyze five or more variables at once. The problem with working with high-dimensional data (data with several variables) is a well-known one, and it is commonly referred to as the *curse of dimensionality*. Having many variables can be a good thing because we can glean more insights from more data. However, it can also be a challenge because there aren't many techniques that can analyze or visualize several variables at once.

In this chapter, we will cover multivariate analysis techniques that can be used to analyze several variables at once. We will cover the following:

- Implementing cluster analysis on multiple variables using Kmeans
- Choosing the optimal number of K clusters in Kmeans
- Profiling Kmeans clusters
- Implementing **Principal Component Analysis (PCA)** on multiple variables
- Choosing the number of principal components
- Analyzing principal components
- Implementing factor analysis on multiple variables
- Determining the number of factors
- Analyzing the factors

Technical requirements

We will leverage the numpy, pandas, matplotlib, seaborn, sklearn, and factor_analyzer libraries in Python for this chapter. This chapter assumes you have sklearn and factor_analyzer installed. If you do not, you can install them using the following code. The code and notebooks for this chapter are available on GitHub at https://github.com/PacktPublishing/Exploratory-Data-Analysis-with-Python-Cookbook:

```
pip install scikit-learn
pip install factor_analyzer
```

Implementing Cluster Analysis on multiple variables using Kmeans

Clustering involves splitting data points into clusters or groups. The goal is to ensure all the data points in a group are similar to each other but dissimilar to data points in other groups. The clusters are determined based on similarity. There are different types of clustering algorithms; we have centroid-based, hierarchical-based, density-based, and distribution-based algorithms. These all have their strengths and the type of data they are best suited for.

In this recipe, we will focus on the most used clustering algorithm, the Kmeans clustering algorithm. This is a centroid-based algorithm that splits data into *K* number of clusters. These clusters are usually predefined by the user before running the algorithm. Each data point is assigned to a cluster based on its distance from the cluster centroid. The goal of the algorithm is to minimize the variance of data points within their corresponding clusters. The process of minimizing the variance is an iterative one.

We will explore how to build a Kmeans model in the sklearn library.

Getting ready

We will work with only two datasets in this chapter – Customer Personality Analysis data from Kaggle and Website Satisfaction Survey data from Kaggle.

Create a folder for this chapter and create a new Python script or Jupyter Notebook file in that folder. Create a data subfolder and place the marketing_campaign.csv and website_survey. csv files in that subfolder. Alternatively, you could retrieve all the files from the GitHub repository.

> **Note**
>
> Kaggle provides the Customer Personality Analysis data for public use data at `https://www.kaggle.com/datasets/imakash3011/customer-personality-analysis`. It also provides the Website Satisfaction Survey data for public use at `https://www.kaggle.com/datasets/hayriyigit/website-satisfaction-survey`.

The *Customer Personality Analysis* data in Kaggle appears in a single-column format, but the data in the repository was transformed into a multiple-column format for easy usage in pandas.

In this chapter, we will use the full datasets of *Customer Personality Analysis* and *Website Satisfaction Survey* for the different recipes. The data is also available in the repository.

How to do it...

We will learn how to perform multivariate analysis using the `sklearn` library:

1. Import the `pandas`, `matplotlib`, `seaborn`, and `sklearn` libraries:

   ```python
   import pandas as pd
   import matplotlib.pyplot as plt
   import seaborn as sns
   from sklearn.preprocessing import StandardScaler
   from sklearn.cluster import KMeans
   ```

2. Load the `.csv` into a dataframe using `read_csv`. Subset the dataframe to include only relevant columns:

   ```python
   marketing_data = pd.read_csv("data/marketing_campaign.csv")
   marketing_data = marketing_data[['MntWines','MntFruits', 'MntMea
   tProducts','MntFishProducts','MntSweetProducts','MntGoldProds',
   'NumDealsPurchases','NumWebPurchases', 'NumCatalogPurchases',
   'NumStorePurchases', 'NumWebVisitsMonth']]
   ```

3. Check the first five rows using the `head` method. Check the number of columns and rows as well as the data types:

   ```python
   marketing_data.head()
   ```

	MntWines	MntFruits	...	NumWebVisitsMonth
0	635	88	...	7
1	11	1	...	5
2	426	49	...	4
3	11	4	...	6
4	173	43	...	5

```
marketing_data.shape
(2240, 11)

marketing_data.dtypes
MntWines                    int64
MntFruits                   int64
MntMeatProducts             int64
...              ...
NumStorePurchases           int64
NumWebVisitsMonth           int64
```

4. Check for missing values using the `isnull` method in pandas. We will use the chaining method to call the `isnull` and `sum` methods on the dataframe in one statement:

```
marketing_data.isnull().sum()
MntWines                    0
MntFruits                   0
...              ...
NumStorePurchases           0
NumWebVisitsMonth           0
```

5. Remove missing values using the `dropna` method. The `inplace` parameter modifies the original object directly without having to create a new object:

```
marketing_data.dropna(inplace=True)

marketing_data.shape
(2240, 11)
```

6. Scale the data using the `StandardScaler` class in the `sklearn` library:

```
scaler = StandardScaler()
marketing_data_scaled = scaler.fit_transform(marketing_data)
```

7. Build the Kmeans model using the Kmeans module in the `sklearn` library:

```
kmeans = KMeans(n_clusters= 4, init='k-means++',random_state= 1)
kmeans.fit(marketing_data_scaled)
```

8. Visualize the Kmeans clusters using `matplotlib`:

```
label = kmeans.fit_predict(marketing_data_scaled)
marketing_data_test = marketing_data.copy()
marketing_data_test['label'] = label
marketing_data_test['label'] = marketing_data_test['label'].
```

```
astype(str)

plt.figure(figsize= (18,10))
sns.scatterplot(data = marketing_data_test , marketing_data_
test['MntWines'], marketing_data_test['MntFruits'], hue =
marketing_data_test['label'])
```

This results in the following output:

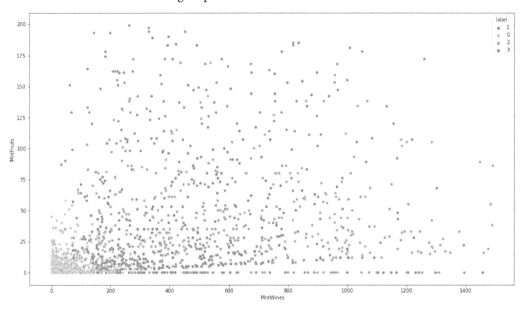

Figure 6.1: A scatterplot of Kmeans clusters across two variables

Well done! We have performed multivariate analysis on our data using kmeans.

How it works...

In this recipe, we used several libraries, such as the pandas, matplotlib, seaborn, and sklearn libraries. In *step 1*, we import all the libraries. In *step 2*, we use the read_csv function to load the .csv file into a pandas dataframe and call it marketing_data. We then subset the dataframe to include only 11 relevant columns. In *step 3*, we get a quick sense of what our data looks like by inspecting the first five rows using the head method. We also get a glimpse of the dataframe shape (the number of rows and columns) and the data types, using the shape and dtypes attributes respectively.

In *step 4*, we check for missing values to ensure there are none; this is a prerequisite for the Kmeans model. We use the isnull and sum methods in pandas to achieve this. This approach is called chaining. We use the chaining method to call multiple methods on an object in one statement. Without chaining, we would have had to call the first method and save the output in a variable before calling the next method on the saved variable. The values displayed from the `isnull` and `sum` chain method indicate the number of missing values per column. Our dataset has no missing values; hence, all the values are zero.

In *step 5*, we use the `dropna` method to remove any missing rows. In this method, we use the `inplace` parameter to modify our existing dataframe without having to create a new dataframe. We also use the `shape` attribute to check the shape of our dataset; this remains the same shape because there are no missing values. In *step 6*, we scale the data using the `StandardScaler` class. Scaling ensures all variables have the same magnitude; it is a critical step for distance-based algorithms such as kmeans. In *step 7*, we build the kmeans model using the `KMeans` class. In the class, we use the `n_clusters` parameter to specify the number of clusters, the `init` parameter to indicate the specific model (kmeans++), and `random_state` to specify a value that ensures reproducibility. kmeans++ optimizes the selection process of the cluster centroid and ensures better-formed clusters. We then use the `fit` method to fit the model to our dataset.

In *step 8*, we use the `fit_predict` method to create our clusters and assign them to a new column called `label`. Using the `scatterplot` function in `seaborn`, we plot the scatterplot of two variables and the cluster label to inspect differences in the clusters. We assign the labels to the `hue` parameter of the scatterplot to distinguish between clusters by using different colors per cluster. The output gives us a sense of how the data points in each cluster are distributed.

There is more...

Kmeans has a few drawbacks. First, it doesn't work too well with categorical variables; it only works well with numerical variables. Second, it is sensitive to outliers, and this can make the output unreliable. Third, the number of *K* clusters needs to be predefined by the user. To solve the categorical data challenge, we can explore the *K*-Prototype algorithm, which is designed to handle both numerical and categorical variables. For outliers, it is advisable to treat outliers before using the algorithm. Alternatively, we can use a density-based algorithm such as DBSCAN, which works well with outliers. In the next recipe, we will cover a technique on how to choose the optimal number of *K* clusters.

Beyond EDA, cluster analysis is also very useful in machine learning. A common use case is customer segmentation, where we need to group data points together based on their similarity (in this case, customers). It can also be used for image classification to group similar images together.

See also...

You can check out the following insightful articles on Kmeans and clustering algorithms:

- https://www.analyticsvidhya.com/blog/2019/08/comprehensive-guide-k-means-clustering/

- https://www.freecodecamp.org/news/8-clustering-algorithms-in-machine-learning-that-all-data-scientists-should-know/

Choosing the optimal number of clusters in Kmeans

One of the major drawbacks of the Kmeans clustering algorithm is the fact that the K number of clusters must be predefined by the user. One of the commonly used techniques to solve this problem is the elbow method. The elbow method uses the **Within Cluster Sum of Squares** (**WCSS**), also called inertia, to find the optimal number of clusters (K). WCSS indicates the total variance within clusters. It is calculated by finding the distance between each data point in a cluster and the corresponding cluster centroid and summing up these distances together.

The elbow method computes the Kmeans for a range of predefined K values – for example, 2–10 – and plots a graph, with the x axis being the number of K clusters and the y axis being the corresponding WCSS for each K cluster.

In this recipe, we will explore how to use the elbow method to identify the optimal number of K clusters. We will use some custom code alongside `sklearn`.

Getting ready

We will work with the Customer Personality Analysis data from Kaggle on this recipe. You can retrieve all the files from the GitHub repository.

How to do it...

We will learn how to choose the optimal number of K clusters using the `sklearn` library:

1. Import the `pandas`, `matplotlib`, `seaborn`, and `sklearn` libraries:

    ```
    import pandas as pd
    import matplotlib.pyplot as plt
    import seaborn as sns
    from sklearn.preprocessing import StandardScaler
    from sklearn.cluster import KMeans
    ```

2. Load the `.csv` into a dataframe using `read_csv`. Subset the dataframe to include only relevant columns:

    ```
    marketing_data = pd.read_csv("data/marketing_campaign.csv")

    marketing_data = marketing_data[['MntWines','MntFruits',
    'MntMeatProducts','MntFishProducts','MntSweetProducts',
    'MntGoldProds','NumDealsPurchases','NumWebPurchases', 'Num
    CatalogPurchases', 'NumStorePurchases', 'NumWebVisitsMonth']]
    ```

3. Check the first five rows using the `head` method. Check the number of columns and rows as well as the data types:

```
marketing_data.head()
    MntWines    MntFruits    ...    NumWebVisitsMonth
0      635          88        ...           7
1       11           1        ...           5
2      426          49        ...           4
3       11           4        ...           6
4      173          43        ...           5


marketing_data.shape
(2240, 11)


marketing_data.dtypes
MntWines                    int64
MntFruits                   int64
MntMeatProducts             int64
...                 ...
NumStorePurchases           int64
NumWebVisitsMonth           int64
```

4. Check for missing values using the `isnull` method in pandas:

```
marketing_data.isnull().sum()
MntWines                    0
MntFruits                   0
...                 ...
NumStorePurchases           0
NumWebVisitsMonth           0
```

5. Remove missing values using the `dropna` method:

```
marketing_data.dropna(inplace=True)

marketing_data.shape
(2240, 11)
```

6. Scale the data using the `StandardScaler` module in the `sklearn` library:

```
scaler = StandardScaler()
marketing_data_scaled = scaler.fit_transform(marketing_data)
```

7. Build the Kmeans model using the `Kmeans` module in the `sklearn` library:

```
kmeans = KMeans(n_clusters= 4, init='k-means++',random_state= 1)
kmeans.fit(marketing_data_scaled)
```

8. Examine the `kmeans` cluster output:

```
label = kmeans.fit_predict(marketing_data_scaled)
marketing_data_output = marketing_data.copy()
marketing_data_output['cluster'] = label
marketing_data_output['cluster'].value_counts()
0    1020
2     475
3     467
1     278
```

9. Find the optimal number of *K* clusters using the elbow technique:

```
distance_values = []
for cluster in range(1,14):
    kmeans = KMeans(n_clusters = cluster, init='k-means++')
    kmeans.fit(marketing_data_scaled)
    distance_values.append(kmeans.inertia_)

cluster_output = pd.DataFrame({'Cluster':range(1,14), 'distance_
values':distance_values})

plt.figure(figsize=(12,6))
plt.plot(cluster_output['Cluster'], cluster_output['distance_
values'], marker='o')
plt.xlabel('Number of clusters')
plt.ylabel('Inertia')
```

This results in the following output:

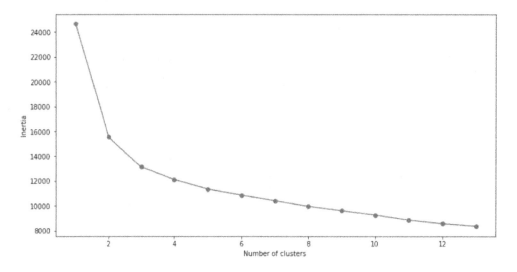

Figure 6.2: A scree plot of inertia and Kmeans clusters

Great! We have identified the optimal number of *K* clusters.

How it works...

In *step 1*, we import all the libraries. In *step 2*, we use `read_csv` to load the `.csv` file into a pandas dataframe and call it `marketing_data`. We then subset the dataframe to include only 11 relevant columns. In *step 3*, we get a quick sense of what our data looks like and its structure, using the `head` method, `shape`, and `dtypes` attributes.

In *step 4* and *step 5*, we check for missing values and drop them if there are any. We use the `isnull` and `dropna` methods respectively.

In *step 6*, we scale the data using the `StandardScaler` class. Scaling ensures that all variables have the same magnitude; this is a critical step for distance-based algorithms such as kmeans. In *step 7*, we build the kmeans model using the `Kmeans` class. In the class, we use the `n_clusters` parameter to specify the number of clusters, the `init` parameter to indicate the specific model (`kmeans++`), and `random_state` to specify a value that ensures reproducibility. `kmeans++` optimizes the selection process of the cluster centroid and ensures better-formed clusters. We then use the `fit` method to fit the model to our dataset.

In *step 8*, we use the `fit_predict` method to create our clusters and assign them to a new variable called `label`. We create a copy of our dataset and assign the `label` variable to a new column in the newly created dataset. We use the `value_counts` method to get a sense of the record count in each cluster. In *step 9*, we create a `for` loop to compute the inertia of the kmeans model, with clusters ranging from 1 to 14. We store the values in a list called `distance_values` and convert the list to a dataframe. We then plot the inertia values against the number of clusters to get our elbow method chart. In the elbow chart, we inspect it to identify where the inertia drops sharply and becomes fairly constant. This happens around clusters 3 to 7. This means we can choose from that range. However, computation cost should also be considered when selecting the optimal *K* cluster because more clusters mean a higher computation cost.

There is more...

Just like the elbow method, silhouette analysis can be used to identify the optimal number of *K* clusters. This method measures the quality of our clusters by checking the distance of samples within a cluster against samples within a neighboring cluster. The score ranges from -1 to 1. A high value indicates that the samples are assigned to the right cluster – that is, samples in a cluster are far from the neighboring clusters. A lower value indicates that the samples have been assigned to the .wrong cluster – that is, samples in a cluster are close to the neighboring clusters. The `silhouette_score` module from `sklearn` can be used for silhouette analysis. Just like the elbow method, we can plot the number of clusters on the *x* axis and the average silhouette score on the *y* axis to visualize our silhouette analysis.

See also...

You can check out the following interesting articles for more on choosing optimal clusters:

- `https://www.analyticsvidhya.com/blog/2021/05/k-mean-getting-the-optimal-number-of-clusters/`

- `https://scikit-learn.org/stable/auto_examples/cluster/plot_kmeans_silhouette_analysis.html`

Profiling Kmeans clusters

Profiling gives us a sense of what each cluster looks like. Through profiling, we can tell the differences and similarities between our various clusters. We can also tell the defining characteristics of each cluster. This is a key step when clustering, especially for exploratory data analysis purposes.

The approach to profiling for numerical fields is to find the mean of the numerical field per cluster. For categorical fields, we can find the percentage occurrence of each category per cluster. The outcome of this computation can then be displayed in various charts, such as tables, boxplots, and scatterplots. A table is typically a good first option because all the values can be displayed at once. Other chart options can then give additional context to the table insights.

We will explore how to profile Kmeans clusters using `pandas`.

Getting ready

We will work with the Customer Personality Analysis data from Kaggle on this recipe. You could retrieve all the files from the GitHub repository.

How to do it...

We will learn how to profile Kmeans clusters using the `pandas` library:

1. Import the `numpy`, `pandas`, `matplotlib`, `seaborn`, and `sklearn` libraries:

    ```
    import numpy as np
    import pandas as pd
    import matplotlib.pyplot as plt
    import seaborn as sns
    from sklearn.preprocessing import StandardScaler
    from sklearn.cluster import KMeans
    ```

2. Load the `.csv` into a dataframe using `read_csv`. Subset the dataframe to include only relevant columns:

    ```
    marketing_data = pd.read_csv("data/marketing_campaign.csv")

    marketing_data = marketing_data[['MntWines','MntFruits', 'MntMea
    tProducts','MntFishProducts','MntSweetProducts','MntGoldProds',
    'NumDealsPurchases','NumWebPurchases', 'NumCatalogPurchases',
    'NumStorePurchases', 'NumWebVisitsMonth']]
    ```

3. Check the first five rows using the `head` method. Check the number of columns and rows as well as the data types:

    ```
    marketing_data.head()
       MntWines    MntFruits    ...    NumWebVisitsMonth
    0      635          88       ...           7
    1       11           1       ...           5
    2      426          49       ...           4
    3       11           4       ...           6
    4      173          43       ...           5

    marketing_data.shape
    (2240, 11)

    marketing_data.dtypes
    MntWines                int64
    MntFruits               int64
    ```

```
MntMeatProducts            int64
...                 ...
NumStorePurchases          int64
NumWebVisitsMonth          int64
```

4. Remove missing values using the `dropna` method:

```
marketing_data.dropna(inplace=True)
```

5. Scale the data using the `StandardScaler` module in the `sklearn` library:

```
scaler = StandardScaler()
marketing_data_scaled = scaler.fit_transform(marketing_data)
```

6. Build the Kmeans model using the `Kmeans` module in the `sklearn` library:

```
kmeans = KMeans(n_clusters= 4, init='k-means++',random_state= 1)
kmeans.fit(marketing_data_scaled)
```

7. Examine the `kmeans` cluster output:

```
label = kmeans.fit_predict(marketing_data_scaled)
marketing_data_output = marketing_data.copy()
marketing_data_output['cluster'] = label
marketing_data_output['cluster'].value_counts()


0    1020
2     475
3     467
1     278
```

8. Get the overall mean per variable to profile the clusters:

```
cols  =['MntWines', 'MntFruits', 'MntMeatProducts',
'MntFishProducts','MntSweetProducts', 'MntGoldProds',
'NumDealsPurchases','NumWebPurchases',
'NumCatalogPurchases','NumStorePurchases','NumWebVisitsMonth']

overall_mean = marketing_data_output[cols].apply(np.mean).T

overall_mean = pd.DataFrame(overall_mean,columns =['overall_
average'])
overall_mean
```

This results in the following output:

	overall_average
MntWines	303.935714
MntFruits	26.302232
MntMeatProducts	166.950000
MntFishProducts	37.525446
MntSweetProducts	27.062946
MntGoldProds	44.021875
NumDealsPurchases	2.325000
NumWebPurchases	4.084821
NumCatalogPurchases	2.662054
NumStorePurchases	5.790179
NumWebVisitsMonth	5.316518

Figure 6.3: The overall average of clusters

9. Get the mean per cluster per variable to profile the clusters:

```
cluster_mean = marketing_data_output.groupby('cluster')[cols].
mean().T
cluster_mean
```

This results in the following output:

cluster	0	1	2	3
MntWines	40.580392	535.892086	627.526316	411.929336
MntFruits	4.913725	98.348921	40.991579	15.188437
MntMeatProducts	21.498039	460.676259	363.021053	110.357602
MntFishProducts	7.219608	133.233813	63.473684	20.351178
MntSweetProducts	5.066667	103.719424	40.835789	15.464668
MntGoldProds	14.696078	98.370504	61.261053	58.186296
NumDealsPurchases	1.869608	1.438849	1.677895	4.505353
NumWebPurchases	2.017647	5.636691	5.277895	6.462527
NumCatalogPurchases	0.556863	5.683453	5.635789	2.436831
NumStorePurchases	3.228431	8.241007	8.713684	6.952891
NumWebVisitsMonth	6.290196	2.920863	3.298947	6.668094

Figure 6.4: The mean per cluster for various variables

10. Concatenate the two datasets to get the final profile output:

```
pd.concat([cluster_mean,overall_mean],axis =1)
```

This results in the following output:

	0	1	2	3	overall_average
MntWines	40.580392	535.892086	627.526316	411.929336	303.935714
MntFruits	4.913725	98.348921	40.991579	15.188437	26.302232
MntMeatProducts	21.498039	460.676259	363.021053	110.357602	166.950000
MntFishProducts	7.219608	133.233813	63.473684	20.351178	37.525446
MntSweetProducts	5.066667	103.719424	40.835789	15.464668	27.062946
MntGoldProds	14.696078	98.370504	61.261053	58.186296	44.021875
NumDealsPurchases	1.869608	1.438849	1.677895	4.505353	2.325000
NumWebPurchases	2.017647	5.636691	5.277895	6.462527	4.084821
NumCatalogPurchases	0.556863	5.683453	5.635789	2.436831	2.662054
NumStorePurchases	3.228431	8.241007	8.713684	6.952891	5.790179
NumWebVisitsMonth	6.290196	2.920863	3.298947	6.668094	5.316518

Figure 6.5: The summarized output

Awesome! We just profiled the clusters defined by our Kmeans model.

How it works...

In *step 1*, we import all the libraries. In *step 2*, we use `read_csv` to load the `.csv` file into a pandas dataframe and call it `marketing_data`. We then subset the dataframe to include only 11 relevant columns. In *step 3*, we get a quick sense of what our data looks like and its structure, using the `method`, `shape`, and `dtypes` attributes.

In *step 4* to *step 5*, we remove missing values because this is a requirement for scaling and the Kmeans model. We then scale the data using the `StandardScaler` class. Scaling ensures that all variables have the same magnitude; this is a critical step for distance-based algorithms such as kmeans. In *step 6*, we build the kmeans model using the `Kmeans` class. In the class, we use the `n_clusters` parameter to specify the number of clusters, the `init` parameter to indicate the specific model (kmeans++), and `random_state` to specify a value that ensures reproducibility. kmeans++ optimizes the selection process of the cluster centroid and ensures better-formed clusters. We then use the `fit` method to fit the model to our dataset.

In *step 7*, we use the `fit_predict` method to create our clusters and assign them to a new variable called `label`. We create a copy of the dataset and assign the `label` variable to a new column in the new dataset. We use the `value_counts` method to get a sense of the count of records in each cluster. In *step 8*, we store the relevant columns that were input into our kmeans model in a list. We subset our dataframe using the list of columns, use the `apply` method to apply the `numpy mean` function to all the rows, and use the `T` method to transpose the data. This gives us the mean per column (average value per column), which will be our baseline for comparing to the clusters. In *step 9*, we find the mean per column per cluster. We use the `groupby` method to group by clusters and the `pandas mean` and `T` methods to achieve this.

In *step 10*, we concatenate the two outputs using the `concat` function. We then review the output by comparing the averages per cluster to each other and the overall average. This will give us a sense of the similarities and differences between the clusters. This review can provide insights into how to name the clusters into more user-friendly or business names. From the profiling, we can deduce the following:

- *Cluster 0* has a very low spend across all categories. It has web visits that are higher than average; however, this doesn't translate into web purchases.

- *Clusters 1 and 2* have a very high spend across product categories and shop primarily at stores and on the web.

- *Cluster 3* has a low spend across the product categories (but close to average). They have the most web visits, and this translates into web purchases. They also have above average store purchases.

There's more...

Profiling can also be done using a supervised machine learning model such as decision trees. The decision trees model helps us to associate a set of rules that are required to arrive at each cluster. The set of rules provided for each cluster displays the defining characteristics of the clusters. Decision trees are very intuitive and easy to explain. The tree can also be visualized to give transparency into rules that have been set. With profiling, the cluster label is used as the target field, while the other variables are used as the model input.

Implementing principal component analysis on multiple variables

Principal Components Analysis (PCA) is a popular dimensionality reduction method that is used to reduce the dimension of very large datasets. It does this by combining multiple variables into new variables called principal components. These components are typically independent of each other and contain valuable information from the original variables.

Even though PCA provides a simple way to analyze large datasets, accuracy is a trade-off. PCA doesn't provide an exact representation of the original data, but it tries to preserve as much valuable information as possible. This means that, most times, it produces an output close enough for us to glean insights from.

Now, we will explore how to implement PCA using the `sklearn` library.

Getting ready

We will work with the Customer Personality Analysis data from Kaggle on this recipe. You can retrieve all the files from the GitHub repository.

How to do it...

We will learn how to implement PCA using the `sklearn` library:

1. Import the `pandas`, `matplotlib`, `seaborn`, and `sklearn` libraries:

```
import pandas as pd
import matplotlib.pyplot as plt
import seaborn as sns
from sklearn.preprocessing import StandardScaler
from sklearn.decomposition import PCA
```

2. Load the `.csv` into a dataframe using `read_csv`. Subset the dataframe to include only relevant columns:

```
marketing_data = pd.read_csv("data/marketing_campaign.csv")


marketing_data = marketing_data[['MntWines','MntFruits', 'MntMea
tProducts','MntFishProducts','MntSweetProducts','MntGoldProds',
'NumDealsPurchases','NumWebPurchases', 'NumCatalogPurchases',
'NumStorePurchases', 'NumWebVisitsMonth']]
```

3. Check the first five rows using the `head` method. Check the number of columns and rows as well as the data types:

```
marketing_data.head()
```

	MntWines	MntFruits	...	NumWebVisitsMonth
0	635	88	...	7
1	11	1	...	5
2	426	49	...	4
3	11	4	...	6
4	173	43	...	5

```
marketing_data.shape
(2240, 11)

marketing_data.dtypes
MntWines                int64
MntFruits               int64
MntMeatProducts         int64

...             ...

NumStorePurchases       int64
NumWebVisitsMonth       int64
```

4. Check for missing values using the `isnull` method in pandas:

```
marketing_data.isnull().sum()
MntWines                0
MntFruits               0

...             ...

NumStorePurchases       0
NumWebVisitsMonth       0
```

5. Remove missing values using the `dropna` method:

```
marketing_data.dropna(inplace=True)

marketing_data.shape
(2240, 11)
```

6. Scale the data using the `StandardScaler` module in the `sklearn` library:

```
x = marketing_data.values
marketing_data_scaled = StandardScaler().fit_transform(x)
```

7. Apply PCA to the dataset using the `PCA` class in the `sklearn` library:

```
pca_marketing = PCA(n_components=6,random_state = 1)
principalComponents_marketing = pca_marketing.fit_
transform(marketing_data_scaled)
```

8. Save the output into a dataframe and Inspect:

```
principal_marketing_data = pd.DataFrame(data =
principalComponents_marketing, columns = ['principal component
1', 'principal component 2',
'principal component 3','principal component 4'
,'principal component 5','principal component 6'])
principal_marketing_data
```

This results in the following output:

	principal component 1	principal component 2	principal component 3	principal component 4	principal component 5	principal component 6
0	3.800461	0.572973	1.254630	1.083547	0.274886	2.368660
1	-2.175610	-0.928702	-0.117578	0.292224	0.323580	-0.105413
2	1.501507	0.123894	0.096791	-0.992810	-1.071276	-0.602728
3	-2.016701	-0.518668	0.025703	0.070743	-0.181590	-0.227872
4	-0.044173	0.763401	0.238572	1.149119	-0.334696	-0.495866
...	...	...	...	...	...	...
2235	2.660651	1.308848	2.151732	-2.178308	1.453732	0.526411
2236	-1.063664	2.738997	-0.463307	0.821222	-0.336517	-0.070777
2237	1.130411	0.004491	-1.519866	-0.539346	-0.947537	-0.493189
2238	1.749883	0.079894	-0.509966	-0.305776	-0.035570	-0.746510
2239	-1.796636	0.271300	0.011374	0.207581	0.083997	0.072001

2240 rows × 6 columns

Figure 6.6: The PCA output dataframe

Great! We have implemented PCA on our dataset.

How it works...

In this recipe, we use several libraries such as the pandas, matplotlib, seaborn, and sklearn libraries. In *step 1*, we import all the libraries. In *step 2*, we use read_csv to load the .csv file into a pandas dataframe and call it marketing_data. We then subset the dataframe to include only 11 relevant columns. In *step 3*, we use the head method to get a view of the first 5 rows of the dataset. We then use the shape attribute to get a sense of the number of rows and columns. We use the dtypes attribute to check the datatypes.

In *step 4*, we check for missing values to ensure there are none; this is a prerequisite for PCA. We use the isnull and sum methods in pandas to achieve this. The values displayed indicate the number of missing values per column. Our dataset has no missing values; hence, all the values are zero.

In *step 5*, we use the dropna method to remove any missing rows. In this method, we use the inplace parameter to modify our existing dataframe. We also use the shape attribute to check the shape of our dataset; this remains the same shape because there are no missing values. In *step 6*, we scale the data using the StandardScaler class. Scaling ensures all variables have the same magnitude; it is a critical step PCA. In *step 7*, we implement PCA using the PCA class. In the class, we use the n_components parameter to specify the number of components and random_state to specify a value that ensures reproducibility. We then use the fit_transform method to fit the model to our dataset.

In *step 8*, we use the PCA output into a dataframe to inspect it.

There is more...

PCA is also commonly used in machine learning for dimensionality reduction. It is used to transform a large dataset with many variables into one with a smaller set of uncorrelated variables, while retaining information from the original data.

This not only simplifies the dataset but also improves the speed and accuracy of machine learning models.

See also...

You can check out the following insightful article:

https://builtin.com/data-science/step-step-explanation-principal-component-analysis

Choosing the number of principal components

Not all principal components provide valuable information. This means we don't need to keep all components to get a good representation of our data; we can keep just a few. We can use a **scree plot** to get a sense of the most useful components. The scree plot plots the components against the proportion of explained variance of each component – that is, the amount of information each component holds in relation to the original variables. The proportion of explained variance is the variance of each component over the sum of the variance from all components. Typically, the higher, the better because a higher proportion means the component will provide a good representation of the original variables. A cumulative explained variance of about 75% is a good target to aim for.

The variance of a component is derived from the eigenvalue of that component. In simple terms, the eigenvalues give us a sense of how much variance (information) in our variables is explained by a component.

We will explore how to identify the optimal number of components using `sklearn` and `matplotlib`.

Getting ready

We will work with the *Customer Personality Analysis* data from Kaggle on this recipe. You can retrieve all the files from the GitHub repository.

How to do it...

We will learn how to perform multivariate analysis using the `sklearn` library:

1. Import the numpy, `pandas`, `matplotlib`, `seaborn`, and `sklearn` libraries:

    ```
    import numpy as np
    import pandas as pd
    ```

```
import matplotlib.pyplot as plt
import seaborn as sns
from sklearn.preprocessing import StandardScaler
from sklearn.decomposition import PCA
```

2. Load the `.csv` file into a dataframe using `read_csv`. Subset the dataframe to include only relevant columns:

```
marketing_data = pd.read_csv("data/marketing_campaign.csv")

marketing_data = marketing_data[['MntWines','MntFruits',
'MntMeat
Products','MntFishProducts','MntSweetProducts','MntGoldProds',
'NumDealsPurchases','NumWebPurchases', 'NumCatalogPurchases',
'NumStorePurchases', 'NumWebVisitsMonth']]
```

3. Check the first five rows using the `head` method. Check the number of columns and rows as well as the data types:

```
marketing_data.head()
     MntWines    MntFruits    ...    NumWebVisitsMonth
0       635          88       ...          7
1        11           1       ...          5
2       426          49       ...          4
3        11           4       ...          6
4       173          43       ...          5

marketing_data.shape
(2240, 11)

marketing_data.dtypes
MntWines               int64
MntFruits              int64
MntMeatProducts        int64
...            ...
NumStorePurchases      int64
NumWebVisitsMonth      int64
```

4. Check for missing values using the `isnull` method in pandas:

```
marketing_data.isnull().sum()
MntWines               0
MntFruits              0
...            ...
NumStorePurchases      0
NumWebVisitsMonth      0
```

5. Remove missing values using the `dropna` method:

    ```
    marketing_data.dropna(inplace=True)

    marketing_data.shape
    (2240, 11)
    ```

6. Scale the data using the `StandardScaler` module in the `sklearn` library:

    ```
    x = marketing_data.values
    marketing_data_scaled = StandardScaler().fit_transform(x)
    ```

7. Apply PCA to the dataset using the `pca` module in the `sklearn` library:

    ```
    pca_marketing = PCA(n_components=6,random_state = 1)
    principalComponents_marketing = pca_marketing.fit_
    transform(marketing_data_scaled)
    ```

8. Check the explained variance for each component:

    ```
    for i in range(0,len(pca_marketing.explained_variance_ratio_)):
        print("Component ",i ,"",pca_marketing.explained_variance_
    ratio_[i])
    ```

    ```
    Component  0    0.46456652843636387
    Component  1    0.1405246545704046
    Component  2    0.07516844380951325
    Component  3    0.06144172878159457
    Component  4    0.05714631700947585
    Component  5    0.047436409149406174
    ```

9. Create a scree plot to check the number of optimal components:

    ```
    plt.figure(figsize= (12,6))

    PC_values = np.arange(pca_marketing.n_components_) + 1
    cummulative_variance = np.cumsum(pca_marketing.explained_
    variance_ratio_)
    plt.plot(PC_values, cummulative_variance, 'o-', linewidth=2,
    color='blue')
    plt.title('Scree Plot')
    plt.xlabel('Principal Components')
    plt.ylabel('Cummulative Explained Variance')
    plt.show()
    ```

This results in the following output:

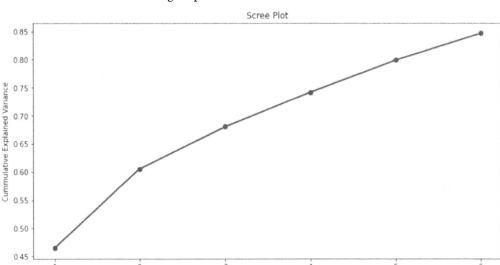

Figure 6.7: A scree plot of cumulative explained variance and principal components

Great. We have identified the optimal number of components for our PCA model.

How it works...

In *step 1*, we import all the libraries. In *step 2*, we use read_csv to load the .csv file into a pandas dataframe and call it marketing_data. We then subset the dataframe to include only 11 relevant columns. In *step 3*, we get a quick sense of what our data looks like and its structure, using the head method, shape, and dtypes attributes.

In *step 4* and *step 5*, we check for missing values and drop them if there are any. We use the isnull and dropna methods respectively.

In *step 6*, we scale the data using the StandardScaler class. Scaling ensures all variables have the same magnitude; it is a critical step for PCA. In *step 7*, we implement PCA on our dataset using the PCA class. In the class, we use the n_components parameter to specify the number of components, and we use random_state to specify a value that ensures reproducibility. We then use the fit_transform method to fit the model to our dataset.

In *step 8*, we create a for loop to print out the explained variance for each component. The explained variance is stored in the explained_variance_ratio_ attribute. In *step 9*, we create a scree plot of the cumulative explained variance. To achieve this, we use the numpy arange function to achieve an evenly spaced array of the range of components (one to six). We apply the cumsum function to the explained variance ratio to calculate the cumulative explained variance ratio of the components. We then use the `plot` function to create a scree plot with the array of components and the corresponding cumulative explained variance as the *x* and *y* axes respectively. When inspecting the scree plot, we usually select the number of components that give a cumulative explained variance of at least 75%. In the scree plot, this is between four and six components.

Analyzing principal components

Principal components are constructed as linear combinations of original variables, which makes them less interpretable and devoid of inherent meaning. This means that after implementing PCA, we need to determine the meaning of the components. One approach to do this is analyzing the relationship between the original variables and the principal components. The values that express this relationship are called loadings.

We will explore how to analyze principal components using `sklearn`.

Getting ready

We will work with the *Customer Personality Analysis* data from Kaggle on this recipe. You can retrieve all the files from the GitHub repository.

How to do it...

We will learn how to analyze the output of a PCA model using the `sklearn` library:

1. Import the `pandas`, `matplotlib`, `seaborn`, and `sklearn` libraries:

    ```
    import pandas as pd
    import matplotlib.pyplot as plt
    import seaborn as sns
    from sklearn.preprocessing import StandardScaler
    from sklearn.decomposition import PCA
    ```

2. Load the `.csv` into a dataframe using `read_csv`. Subset the dataframe to include only relevant columns:

    ```
    marketing_data = pd.read_csv("data/marketing_campaign.csv")

    marketing_data = marketing_data[['MntWines','MntFruits', 'MntMea
    tProducts','MntFishProducts','MntSweetProducts','MntGoldProds',
    'NumDealsPurchases','NumWebPurchases', 'NumCatalogPurchases',
    'NumStorePurchases', 'NumWebVisitsMonth']]
    ```

3. Check the first five rows using the `head` method. Check the number of columns and rows as well as the data types:

```
marketing_data.head()
    MntWines    MntFruits    ...    NumWebVisitsMonth
0       635          88       ...            7
1        11           1       ...            5
2       426          49       ...            4
3        11           4       ...            6
4       173          43       ...            5

marketing_data.shape
(2240, 11)

marketing_data.dtypes
MntWines                    int64
MntFruits                   int64
MntMeatProducts             int64
...           ...
NumStorePurchases           int64
NumWebVisitsMonth           int64
```

4. Check for missing values using the `isnull` method in pandas:

```
marketing_data.isnull().sum()
MntWines                    0
MntFruits                   0
...           ...
NumStorePurchases           0
NumWebVisitsMonth           0
```

5. Remove missing values using the `dropna` method:

```
marketing_data.dropna(inplace=True)

marketing_data.shape
(2240, 11)
```

6. Scale the data using the `StandardScaler` class in the `sklearn` library:

```
x = marketing_data.values
marketing_data_scaled = StandardScaler().fit_transform(x)
```

7. Apply PCA to the dataset using the `pca` module in the `sklearn` library:

```
pca_marketing = PCA(n_components=6,random_state = 1)
principalComponents_marketing = pca_marketing.fit_
transform(marketing_data_scaled)
```

8. Extract the loadings and examine them:

```
loadings_df = pd.DataFrame(pca_marketing.components_).T
loadings_df = loadings_df.set_index(marketing_data.columns)
loadings_df
```

This results in the following output:

	0	1	2	3	4	5
MntWines	0.327941	0.222837	-0.435535	-0.208662	-0.087749	0.243052
MntFruits	0.323026	-0.130151	0.376355	0.140996	-0.224386	-0.012065
MntMeatProducts	0.354452	-0.130388	-0.209744	0.305524	0.151587	0.354552
MntFishProducts	0.333163	-0.142444	0.345355	0.150907	-0.049328	0.050934
MntSweetProducts	0.321179	-0.104676	0.363038	0.115690	-0.350306	0.047819
MntGoldProds	0.265813	0.189065	0.405995	-0.416516	0.693513	-0.128306
NumDealsPurchases	-0.042299	0.636331	0.077169	0.661013	0.144609	-0.268801
NumWebPurchases	0.245131	0.493262	0.039387	-0.358028	-0.270322	0.161445
NumCatalogPurchases	0.360813	0.009298	-0.269517	0.235563	0.316932	0.252435
NumStorePurchases	0.329634	0.187143	-0.241080	-0.112152	-0.297203	-0.574865
NumWebVisitsMonth	-0.277380	0.407525	0.265537	-0.020082	-0.173500	0.548833

Figure 6.8: The PCA dataframe for the loadings

9. Filter out loadings below a specific threshold:

```
loadings_df.where(abs(loadings_df) >= 0.35)
```

This results in the following output:

	0	1	2	3	4	5
MntWines	NaN	NaN	-0.435535	NaN	NaN	NaN
MntFruits	NaN	NaN	0.376355	NaN	NaN	NaN
MntMeatProducts	0.354452	NaN	NaN	NaN	NaN	0.354552
MntFishProducts	NaN	NaN	NaN	NaN	NaN	NaN
MntSweetProducts	NaN	NaN	0.363038	NaN	-0.350306	NaN
MntGoldProds	NaN	NaN	0.405995	-0.416516	0.693513	NaN
NumDealsPurchases	NaN	0.636331	NaN	0.661013	NaN	NaN
NumWebPurchases	NaN	0.493262	NaN	-0.358028	NaN	NaN
NumCatalogPurchases	0.360813	NaN	NaN	NaN	NaN	NaN
NumStorePurchases	NaN	NaN	NaN	NaN	NaN	-0.574865
NumWebVisitsMonth	NaN	0.407525	NaN	NaN	NaN	0.548833

Figure 6.9: The PCA output for various thresholds

Great! We have analyzed the components from our PCA model.

How it works...

In *step 1*, we import all the libraries. In *step 2*, we use read_csv to load the .csv file into a pandas dataframe and call it marketing_data. We then subset the dataframe to include only 11 relevant columns. In *step 3*, we get a quick sense of what our data looks like and its structure using the head method, shape and dtypes attributes.

In *step 4* and *step 5*, we check for missing values and drop them if there are any. We use the isnull and dropna methods respectively.

In *step 6*, we scale the data using the StandardScaler class. Scaling ensures all variables have the same magnitude; it is a critical step for PCA. In *step 7*, we implement PCA on our dataset using the PCA class. In the class, we use the n_components parameter to specify the number of components, and we use random_state to specify a value that ensures reproducibility. We then use the fit_transform method to fit the model to our dataset.

In *step 8*, we extract the loadings from the n_components_ attribute and store them in a dataframe. We then transpose the dataframe using the T method. We set the index of the dataframe to the column names using the set_index method.

In *step 9*, we filter the loadings by a threshold value to make it easier to interpret and understand. The output shows us how the variables are related to the various components.

There's more...

A common setback of the loadings is the fact that several variables can be correlated to a single component, thereby making it less interpretable. A common solution to this is called **Rotation in proper case**.

Rotation rotates the components into a structure that makes them more interpretable. Rotation minimizes the number of variables that are strongly correlated with components and attempts to associate each variable to only one component, aiding interpretability. The outcome of rotation is a large correlation between each component and a specific set of variables and a negligible correlation with other variables.

Currently, `sklearn` doesn't have an implementation of rotation. Therefore, custom code will be required to achieve it.

See also...

Python has an implementation for rotation in the `factor_analyzer` library for factor analysis. You can check the *Implementing Factor Analysis on multiple variables* recipe to get a sense of what a rotated output looks like.

Implementing factor analysis on multiple variables

Just like PCA, **factor analysis** can be used for dimensionality reduction. It can be used to condense multiple variables into a smaller set of variables called factors that are easier to analyze and understand. A factor is a latent or hidden variable that describes the relationship of observed variables (i.e., variables captured in our dataset). The key concept is that multiple variables in our dataset have similar responses because they are associated with a specific theme or hidden variable that is not directly measured. For example, responses to variables such as the taste of food, food temperature, and freshness of food are likely to be similar because they have a common theme (factor), which is food quality. Factor analysis is quite popular in the analysis of survey data.

In this recipe, we will explore how to apply factor analysis to a dataset using the `factor_analyzer` library.

Getting ready

We will work with the *Website Survey* data from Kaggle on this recipe. You can retrieve all the files from the GitHub repository. Each of the variables (q1 to q26) in the Website Survey data is tied to a survey question. The survey questions tied to each variable can be seen in the metadata of the Survey data on Kaggle.

> **Note**
> Kaggle provides the Website Satisfaction Survey data for public use at `https://www.kaggle.com/datasets/hayriyigit/website-satisfaction-survey`.

How to do it...

We will learn how to perform factor analysis using the `factor_analyzer` library:

1. Import the `pandas`, `matplotlib`, `seaborn`, `sklearn`, and `factor_analyzer` libraries:

    ```
    import pandas as pd
    import matplotlib.pyplot as plt
    import seaborn as sns
    from sklearn.preprocessing import StandardScaler
    from factor_analyzer import FactorAnalyzer
    from factor_analyzer.factor_analyzer import calculate_kmo
    ```

2. Load the `.csv` into a dataframe using `read_csv`. Subset the dataframe to include only relevant columns:

    ```
    satisfaction_data = pd.read_csv("data/website_survey.csv")

    satisfaction_data = satisfaction_data[['q1',
    'q2', 'q3','q4', 'q5', 'q6', 'q7', 'q8',
    'q9',                                    'q10','q11','q12',
    'q13', 'q14','q15',
    'q16', 'q17', 'q18', 'q19', 'q20', 'q21', 'q22', 'q23',
    'q24','q25', 'q26']]
    ```

3. Check the first five rows using the `head` method. Check the number of columns and rows as well as the data types:

    ```
    satisfaction_data.head()
         q1    q2    q3    q4   ...    q25    q26
    0     9     7     6     6   ...      5      3
    1    10    10    10     9   ...      9      8
    2    10    10    10    10   ...      8      8
    3     5     8     5     5   ...     10      6
    4     9    10     9    10   ...     10     10

    satisfaction_data.shape
    (73, 26)
    ```

```
Satisfaction_data.dtypes
q1        int64
q2        int64
q3        int64
q4        int64
...       ...
q25       int64
q26       int64
```

4. Check for missing values using the `isnull` method in pandas:

```
satisfaction_data.isnull().sum()
q1                  0
q2                  0
q3                  0
q4                  0
...                 ...
q25                 0
q26                 0
```

5. Remove missing values using the `dropna` method:

```
satisfaction_data.dropna(inplace=True)

satisfaction_data.shape
(73, 26)
```

6. Check for multicollinearity using the `corr` method in pandas:

```
satisfaction_data.corr()[(satisfaction_data.corr()>0.9) &
(satisfaction_data.corr()<1)]

     q1   q2   q3   q4   ...   q25   q26
q1
q2
q3
q4
...
q25
q26
```

7. Test the suitability of the data using the `calculate_kmo` class in the `factor_analyzer` library:

```
kmo_all,kmo_model=calculate_kmo(satisfaction_data)
kmo_model
0.86476
```

8. Apply factor analysis to the dataset using the `FactorAnalyzer` class in the `factor_analyzer` library:

```
fa = FactorAnalyzer(n_factors = 6, rotation="varimax")
fa.fit(satisfaction_data)
```

9. Check the loadings:

```
loadings_output = pd.DataFrame(fa.loadings_,index=satisfaction_data.columns)
loadings_output
```

This results in the following output:

	0	1	2	3	4	5
q1	0.091495	0.773220	-0.075197	0.323028	0.025180	-0.011134
q2	0.170028	0.835743	0.138555	0.107889	0.059724	0.043701
q3	0.030677	0.714850	0.035009	0.096148	0.127674	0.218543
q4	0.129466	0.816115	0.068117	-0.029781	0.133684	0.100296
q5	0.190598	0.651182	0.238826	0.368984	0.036734	0.015868
q6	0.216372	0.121810	0.243359	0.804852	0.108610	0.049316
q7	0.412134	0.238174	0.072134	0.625986	0.165875	0.113483
q8	0.184605	0.288331	0.286474	0.525538	0.206809	0.339109
q9	0.294554	0.006791	0.569362	0.226370	-0.114852	0.530618
q10	0.299583	0.045600	0.639910	0.171549	0.224400	0.157741
q11	0.370816	0.261551	0.503414	0.297665	0.302885	0.183464
q12	0.179816	0.099230	0.891192	0.207311	0.093948	0.033167
q13	0.410334	0.115936	0.298440	0.352660	0.616465	-0.022653
q14	0.404361	0.293690	0.311951	0.188599	0.430506	0.173400
q15	0.232076	0.282038	0.342465	0.573959	0.120769	0.177804
q16	0.405849	0.070595	0.600486	0.137131	0.282366	0.160197
q17	0.414700	0.227079	0.164386	0.534676	0.396478	0.219793
q18	0.389244	0.247077	0.225246	0.078194	0.245236	0.406661
q19	0.456111	0.243858	0.288471	0.224083	0.464653	0.298386
q20	0.418895	0.358063	0.347444	0.326688	0.374696	0.342628
q21	0.221965	0.441209	0.190201	0.336798	0.228637	0.565586
q22	0.475269	0.241734	0.245755	0.418891	0.451944	0.342739
q23	0.757977	0.095174	0.282660	0.252791	0.169766	0.052193
q24	0.805413	0.110427	0.143583	0.193832	0.194302	0.127148
q25	0.667780	0.254834	0.306257	0.203562	0.185318	0.151899
q26	0.768589	0.129073	0.293272	0.265155	0.040168	0.152406

10. Well done! You have performed factor analysis on your dataset.

How it works...

In this recipe, we use several libraries such as pandas, matplotlib, seaborn, and factor_ analyzer libraries. In *step 1*, we import all the libraries. In *step 2*, we use read_csv to load the .csv file into a pandas dataframe and call it marketing_data. We then subset the dataframe to include only 26 relevant columns. In *step 3*, we get a quick sense of what our data looks like by inspecting the first five rows using the head method. We also get a glimpse of the dataframe shape (number of rows and columns) and the data types using the shape and dtypes attributes respectively.

In *step 4*, we check for missing values to ensure there are none; this is a prerequisite. We use the isnull and sum methods in pandas to achieve this. Usually, the values displayed indicate the number of missing values per column. Our dataset has no missing values; hence, all the values are zero.

In *step 5*, we use the dropna method to remove any missing rows. In this method, we use the inplace parameter to modify our existing dataframe. We also use the shape attribute to check the shape of our dataset; this remains the same shape because there are no missing values. In *step 6*, we check for multicollinearity by using the corr method and setting a threshold of greater than 90% to identify columns that are highly correlated. Typically, one of these columns needs to be removed because factor analysis assumes the absence of multicollinearity. The output of this check is empty because there are no columns greater than the threshold.

In *step 7*, we test the suitability of the data for factor analysis using the calculate_kmo class. The **Kaiser-Meyer-Olkin** (**KMO**) test checks for the correlation between our variables; this is a critical prerequisite for a dimensionality reduction technique such as factor analysis. KMO scores range from 0 to 1, and values greater than 0.6 are typically better suited for factor analysis.

In *step 8*, we implement factor analysis on our dataset using the FactorAnalyzer class. In the class, we use the n_factors parameter to specify the number of factors and the rotation parameter to indicate the specific rotation to be used. We then use the fit method to fit the model to our dataset.

In *step 9*, we use the loadings attribute to extract the loadings, and we save this in a dataframe. The loadings indicate the relationship between the original variables and the underlying factors.

There is more...

Just like PCA, factor analysis can also be used in machine learning for dimensionality reduction. When multiple variables are condensed into a smaller set of variables (factors), this simplifies the data, and the output can be used as inputs into a machine learning model to improve speed and accuracy.

Determining the number of factors

The number of factors produced is typically equal to the number of variables in our dataset. However, a significant proportion of the valuable information is contained in only a few factors. This means by keeping only a few factors, we can still get a good representation of our data.

A scree plot can be used to determine the number of factors. The scree plot plots the factors against their eigenvalues. The eigenvalues give us a sense of how much variance (information) in our dataset is explained by the factors. The thumb rule is to select factors whose eigenvalues are greater than 1. Because our variables are typically scaled, the eigenvalue (variance) of a single variable is equal to 1. Hence, useful factors need to explain more information than a single variable, since a factor is meant to be a combination of variables.

We will explore how to check the optimal number of factors using the `factor_analyzer` and `matplotlib` libraries.

Getting ready

We will work with the *Website Survey* data from Kaggle on this recipe. You can retrieve all the files from the GitHub repository.

How to do it...

We will learn how to identify the optimal number of factors using the `factor_analyzer` and `matplotlib` libraries:

1. Import the `numpy`, `pandas`, `matplotlib`, `seaborn`, `sklearn`, and `factor_analyzer` libraries:

```
import numpy as np
import pandas as pd
import matplotlib.pyplot as plt
import seaborn as sns
from sklearn.preprocessing import StandardScaler
from factor_analyzer import FactorAnalyzer
from factor_analyzer.factor_analyzer import calculate_kmo
```

2. Load the `.csv` into a dataframe using `read_csv`. Subset the dataframe to include only relevant columns:

```
satisfaction_data = pd.read_csv("data/website_survey.csv")

satisfaction_data = satisfaction_data[['q1',
'q2', 'q3','q4', 'q5', 'q6', 'q7', 'q8',
'q9',                                      'q10','q11','q12',
'q13', 'q14','q15',
'q16', 'q17', 'q18', 'q19', 'q20', 'q21', 'q22', 'q23',
'q24','q25', 'q26']]
```

3. Check the first five rows using the head method. Check the number of columns and rows as well as the data types:

```
satisfaction_data.head()
      q1    q2    q3    q4    ...    q25    q26
0      9     7     6     6    ...      5      3
1     10    10    10     9    ...      9      8
2     10    10    10    10    ...      8      8
3      5     8     5     5    ...     10      6
4      9    10     9    10    ...     10     10

satisfaction_data.shape
(73, 26)

satisfaction.dtypes
q1        int64
q2        int64
q3        int64
q4        int64
...       ...
q25       int64
q26       int64
```

4. Check for missing values using the isnull method in pandas:

```
satisfaction_data.isnull().sum()
q1              0
q2              0
q3              0
q4              0
...             ...
q25             0
q26             0
```

5. Remove missing values using the dropna method:

```
satisfaction_data.dropna(inplace=True)

satisfaction_data.shape
(73, 26)
```

6. Check for multicollinearity using the `corr` method in `pandas`:

```
satisfaction_data.corr()[(satisfaction_data.corr()>0.9) &
(satisfaction_data.corr()<1)]
```

	q1	q2	q3	q4	...	q25	q26
q1							
q2							
q3							
q4							
...							
q25							
q26							

7. Test the suitability of the data using the `calculate_kmo` class in the `factor_analyzer` library:

```
kmo_all,kmo_model=calculate_kmo(satisfaction_data)
kmo_model
0.86476
```

8. Apply factor analysis to the dataset using the `FactorAnalyzer` class in the `factor_analyzer` library:

```
fa = FactorAnalyzer(n_factors = 6, rotation="varimax")
fa.fit(satisfaction_data)
```

9. Create a scree plot:

```
fa = FactorAnalyzer(rotation = 'varimax',n_factors=marketing_
data.shape[1])
fa.fit(marketing_data)
ev,_ = fa.get_eigenvalues()

factor_values = np.arange(marketing_data.shape[1]) + 1

plt.figure(figsize= (12,6))

plt.plot(factor_values,ev,'o-', linewidth=2)
plt.title('Scree Plot')
plt.xlabel('Factors')
plt.ylabel('Eigen Value')
plt.grid()
```

This results in the following output:

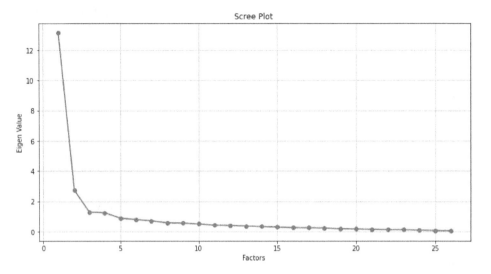

Figure 6.10: A screen plot of the eigenvalues and factors

Awesome! We just found the optimal number of factors for our factor analysis.

How it works...

In *step 1*, we import all the libraries. In *step 2*, we use `read_csv` to load the `.csv` file into a `pandas` dataframe and call it `marketing_data`. We then subset the dataframe to include only 26 relevant columns. In *step 3*, we get a quick sense of what our data looks like and its structure, using the `head` method, `shape`, and `dtypes` attributes.

In *step 4* and *step 5*, we check for missing values and drop missing values if there are any. We use the `isnull` and `dropna` methods respectively. In *step 6* and *step 7*, we test the suitability of the data for factor analysis. We check for multicollinearity and apply the KMO test using the `calculate_kmo` class.

In *step 8*, we implement factor analysis on our dataset using the `FactorAnalyzer` class. In the class, we use the `n_factors` parameter to specify the number of factors and the `rotation` parameter to indicate the specific rotation to be used. We then use the `fit` method to fit the model to our dataset.

In *step 9*, we extract the eigenvalues from the `get_eigenvalues` method. This method stores the original eigenvalues and the common factor eigenvalues. We use the `numpy arange` function to achieve an evenly spaced array of the range of factors (1–26). We then use the `plot` function to create a scree plot with the array of factors and the corresponding eigenvalues as the *x* and *y* axes respectively. Usually, we can select the number of factors that have eigenvalues greater than 1. From the scree plot, we can select five factors.

Analyzing the factors

After generating the factors of our dataset, it would be interesting to understand the contribution of the original variables to the factors. Since factors are a linear combination of the original variables, they are typically less interpretable until they are analyzed. The goal is to understand the information that each factor conveys in order to name them accordingly. There are three concepts that aid the analysis of factors.

The first is the loadings; they express the relationship between the original variables and the underlying factors. In simple terms, it is basically the correlation coefficient between the variables and the underlying factors. The loading values range from -1 to 1, where values closer to -1 or 1 indicate that a factor has a significant influence on the variables.

The second is communality, which displays the proportion of each variable's variance that is explained by the underlying factors.

The third is rotation, which rotates the factors into a structure that makes them more interpretable. Having several variables that are correlated to the factors can make factors less interpretable. Rotation minimizes the number of variables that are strongly correlated with factors and attempts to associate each variable with only one factor, aiding interpretability. The outcome of rotation is a large correlation between each factor and a specific set of variables and a negligible correlation with other variables.

In this recipe, we will explore how to analyze the underlying factors in our dataset using the `factor_analyzer` library.

Getting ready

We will work with the *Website Survey* data from Kaggle on this recipe. You can retrieve all the files from the GitHub repository.

How to do it...

We will learn how to analyze factors using the `factor_analyzer` library:

1. Import the `pandas`, `matplotlib`, `seaborn`, `sklearn`, and `factor_analyzer` libraries:

    ```
    import pandas as pd
    import matplotlib.pyplot as plt
    import seaborn as sns
    from sklearn.preprocessing import StandardScaler
    from factor_analyzer import FactorAnalyzer
    from factor_analyzer.factor_analyzer import calculate_kmo
    ```

2. Load the `.csv` into a dataframe using `read_csv`. Subset the dataframe to include only relevant columns:

```
satisfaction_data = pd.read_csv("data/website_survey.csv")

satisfaction_data = satisfaction_data[['q1',
'q2', 'q3','q4', 'q5', 'q6', 'q7', 'q8',
'q9',                                        'q10','q11','q12',
'q13', 'q14','q15',
'q16', 'q17', 'q18', 'q19', 'q20', 'q21', 'q22', 'q23',
'q24','q25', 'q26']]
```

3. Check the first five rows using the `head` method. Check the number of columns and rows as well as the data types:

```
satisfaction_data.head()
     q1    q2    q3    q4    ...    q25   q26
0    9     7     6     6     ...    5     3
1    10    10    10    9     ...    9     8
2    10    10    10    10    ...    8     8
3    5     8     5     5     ...    10    6
4    9     10    9     10    ...    10    10


satisfaction_data.shape
(73, 26)


satisfaction.dtypes
q1       int64
q2       int64
q3       int64
q4       int64
...      ...
q25      int64
q26      int64
```

4. Check for missing values using the `isnull` and `sum` methods in pandas:

```
satisfaction_data.isnull().sum()
q1              0
q2              0
q3              0
q4              0
...             ...
q25             0
q26             0
```

5. Remove missing values using the dropna method:

```
satisfaction_data.dropna(inplace=True)

satisfaction_data.shape
(73, 26)
```

6. Check for multicollinearity using the corr method in pandas:

```
satisfaction_data.corr()[(satisfaction_data.corr()>0.9) &
(satisfaction_data.corr()<1)]
```

	q1	q2	q3	q4	...	q25	q26
q1							
q2							
q3							
q4							
...							
q25							
q26							

7. Test the suitability of the data using the calculate_kmo class in the factor_ analyzer library:

```
kmo_all,kmo_model=calculate_kmo(satisfaction_data)
kmo_model
0.86476
```

8. Apply factor analysis to the dataset using the FactorAnalyzer class in the factor_ analyzer library:

```
fa = FactorAnalyzer(n_factors = 5, rotation="varimax")
fa.fit(satisfaction_data)
```

9. Generate the loadings:

```
loadings_output = pd.DataFrame(fa.loadings_,index=satisfaction_
data.columns)
loadings_output
```

This results in the following output:

	0	1	2	3	4
q1	0.099961	0.774357	-0.079795	0.321797	0.015576
q2	0.178909	0.836019	0.135454	0.101916	0.063459
q3	0.041438	0.713113	0.055440	0.095694	0.254581
q4	0.146058	0.811110	0.070999	-0.031095	0.166014
q5	0.202457	0.655376	0.237125	0.362648	0.002515
q6	0.239624	0.122927	0.257562	0.802537	0.059995
q7	0.436143	0.237686	0.090813	0.621522	0.151444
q8	0.201393	0.286128	0.332926	0.520492	0.355756
q9	0.253617	0.030815	0.584818	0.208287	0.250758
q10	0.333174	0.045337	0.656804	0.164210	0.182566
q11	0.414787	0.256035	0.521090	0.292189	0.270524
q12	0.197599	0.103579	0.900842	0.194721	-0.009453
q13	0.522679	0.116309	0.293003	0.351973	0.285052
q14	0.473860	0.285857	0.327055	0.193583	0.338894
q15	0.247173	0.283218	0.367144	0.565713	0.166359
q16	0.448919	0.067157	0.611020	0.131054	0.223560
q17	0.477713	0.219711	0.189818	0.533798	0.372763
q18	0.401174	0.244422	0.274638	0.075665	0.416154
q19	0.522615	0.233302	0.317482	0.224103	0.469467
q20	0.464073	0.350758	0.384202	0.321777	0.452556
q21	0.229490	0.436125	0.265986	0.329182	0.533499
q22	0.536084	0.232606	0.285820	0.418030	0.495126
q23	0.782388	0.096663	0.289445	0.239066	0.062135
q24	0.821780	0.110341	0.160566	0.182693	0.148196
q25	0.680231	0.254174	0.326707	0.191116	0.166604
q26	0.730739	0.133893	0.321196	0.248114	0.083717

Figure 6.11: The loadings dataframe for the factors

10. Filter the loadings with a specific threshold:

```
loadings_output.where(abs(loadings_output) > 0.5)
```

	0	1	2	3	4
q1	NaN	0.774357	NaN	NaN	NaN
q2	NaN	0.836019	NaN	NaN	NaN
q3	NaN	0.713113	NaN	NaN	NaN
q4	NaN	0.811110	NaN	NaN	NaN
q5	NaN	0.655376	NaN	NaN	NaN
q6	NaN	NaN	NaN	0.802537	NaN
q7	NaN	NaN	NaN	0.621522	NaN
q8	NaN	NaN	NaN	0.520492	NaN
q9	NaN	NaN	0.584818	NaN	NaN
q10	NaN	NaN	0.656804	NaN	NaN
q11	NaN	NaN	0.521090	NaN	NaN
q12	NaN	NaN	0.900842	NaN	NaN
q13	0.522679	NaN	NaN	NaN	NaN
q14	NaN	NaN	NaN	NaN	NaN
q15	NaN	NaN	NaN	0.565713	NaN
q16	NaN	NaN	0.611020	NaN	NaN
q17	NaN	NaN	NaN	0.533798	NaN
q18	NaN	NaN	NaN	NaN	NaN
q19	0.522615	NaN	NaN	NaN	NaN
q20	NaN	NaN	NaN	NaN	NaN
q21	NaN	NaN	NaN	NaN	0.533499
q22	0.536084	NaN	NaN	NaN	NaN
q23	0.782388	NaN	NaN	NaN	NaN
q24	0.821780	NaN	NaN	NaN	NaN
q25	0.680231	NaN	NaN	NaN	NaN
q26	0.730739	NaN	NaN	NaN	NaN

Figure 6.12: The loadings dataframe with thresholds

11. Generate the communalities:

```
pd.DataFrame(fa.get_communalities(),index=satisfaction_data.
columns,columns=['Communalities'])
```

This results in the following output:

	Communalities
q1	0.719784
q2	0.763698
q3	0.587289
q4	0.712800
q5	0.658255
q6	0.786534
q7	0.664188
q8	0.630742
q9	0.513547
q10	0.604748
q11	0.667696
q12	0.899296
q13	0.577711
q14	0.565546
q15	0.623808
q16	0.646538
q17	0.736405
q18	0.475018
q19	0.698971
q20	0.794353
q21	0.706602
q22	0.843083
q23	0.766267
q24	0.768617
q25	0.698338
q26	0.723643

Figure 6.13: The communalities dataframe

Great! We have now analyzed the factors from our factor analysis.

How it works...

In *step 1*, we import all the libraries. In *step 2*, we use `read_csv` to load the `.csv` file into a `pandas` dataframe and call it `marketing_data`. We then subset the dataframe to include only 26 relevant columns. In *step 3*, we get a quick sense of what our data looks like and its structure, using the `head`, `shape`, and `dtypes` attributes.

In *step 4* and *step 5*, we check for missing values and drop missing values if there are any. We use the `isnull` and `dropna` methods respectively. In *step 6* and *step 7*, we test the suitability of the data for factor analysis. We check for multicollinearity and apply the KMO test using the `calculate_kmo` method.

In *step 8*, we implement factor analysis on our dataset using the `FactorAnalyzer` class. In the class, we use the `n_factors` parameter to specify the number of factors and the `rotation` parameter to indicate the specific rotation to be used. We then use the `fit` method to fit the model to our dataset.

In *step 9* and *step 10*, we generate the loadings and apply a threshold of 0.5 to them. This makes the loadings matrix easier to interpret and understand because we can now see the variables that are strongly correlated with each factor. Applying the threshold helps us to associate each variable with only one factor. In *step 11*, we generate the communalities to see the proportion of each variable's variance that is explained by the underlying factors.

From the loadings, we can see the variables that have the most significant influence on the factors. These are variables with a loading of *greater than 0.5*. For example, for *factor 1*, the q13, q19, q22, q23, q24, q25, and q26 variables have a loading of greater than 0.5 and, therefore, are considered to have a significant influence on factor 1. When we review these variables, we can see that they all seem to be related to the reliability of the website. Therefore, we can call factor 1 reliability. For the other factors, we can go through similar steps and name them appropriately. The outcome of this process gives the following:

- **Reliability (factor 1)**:

    - q13: On this website, everything is consistent

    - q19: Overall, I am satisfied with the interface of this website

    - q22: Overall, I am satisfied with the accuracy of this website related to the buying process

    - q23: I trust the information presented on this website

    - q24: This website is credible for me

    - q25: I would visit this website again

    - q26: I would recommend this website to my friend

- **Interface design (factor 2):**

  - q1: It is easy to read the text on this website with the used font type and size

  - q2: The font color is appealing on this website

  - q3: The text alignment and spacing on this website make the text easy to read

  - q4: The color scheme of this website is appealing

  - q5: The use of color or graphics enhances navigation

- **Personalization (factor 3):**

  - q9: This website provides adequate feedback to assess my progression when I perform a task

  - q10: This website offers customization

  - q11: This website offers versatility in the ordering process

  - q12: This website provides content tailored to an individual

  - q16: It is easy to personalize or narrow the buying process

- **User-friendliness (factor 4):**

  - q6: The information content helps with buying decisions by comparing the information about products or services

  - q7: The information content provided by this website meets my needs

  - q8: Contents and information support for reading and learning about buying process

  - q15: It is obvious where the buying button and links are on this website

  - q16: It is easy to personalize or narrow the buying process

  - q17: It is easy to learn to use the website

- **Speed (factor 5):**

  - q21: Overall, I am satisfied with the amount of time it took to buy a product

7

Analyzing Time Series Data in Python

In this chapter, we will learn how to analyze time-dependent data, also known as time series data. Time series data is data that changes over time; it is a sequence of data points that are recorded over regular time intervals such as hourly, daily, weekly, and so on. When we analyze time series data, we can typically uncover patterns or trends that repeat over time. Time series data has the following components:

- **Trend**: This shows the overall direction of the time series data over time. This can be upward, downward, or flat.

- **Seasonal variations**: These are periodic fluctuations that reoccur over fixed timeframes. In simple terms, seasonality exists when a pattern is influenced by seasonal factors, for example, the quarter of the year or the month of the year. Seasonal patterns typically have a fixed period since they are influenced by seasonal factors.

- **Cyclical variations**: These are fluctuations that happen over time but without a fixed period. This means they don't have a set time interval for repetition. The prosperity and recession within an economy or business are examples of cyclical patterns; these fluctuations cannot be tied to a fixed period. The length of a cycle is not fixed, and it is typically longer than that of seasonal patterns. Also, the magnitude per cyclical pattern varies, unlike the magnitude per seasonal pattern, which is quite consistent.

- **Irregular variation/noise**: These are variations or fluctuations that are unpredictable or unexpected.

These components can be represented as follows:

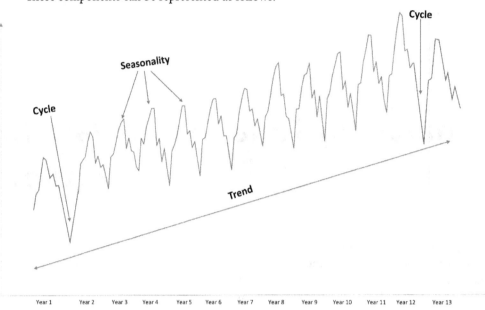

Figure 7.1: Time series data components

In this chapter, we will discuss common techniques for analyzing time series data. The chapter includes the following key topics:

- Using line and box plots to visualize time series data
- Spotting patterns in time series
- Performing time series data decomposition
- Performing smoothing – moving average
- Performing smoothing – exponential smoothing
- Performing stationarity checks on time series data
- Differencing time series data
- Using correlation plots to visualize time series data

Technical requirements

We will leverage the pandas, matplotlib, seaborn, and statsmodels libraries in Python for this chapter. The code and notebooks for this chapter are available on GitHub at https://github.com/PacktPublishing/Exploratory-Data-Analysis-with-Python-Cookbook.

Using line and boxplots to visualize time series data

The line chart connects time series data points through a straight line that displays the peaks (high points) and valleys (low points) in the data. The x axis of the line chart typically represents our time intervals, while the y axis represents a variable of interest that we intend to track in relation to time. With the line chart, it is easy to spot trends or changes over time.

On the other hand, the boxplot gives us a sense of the underlying distribution of a dataset through five key metrics. The metrics include minimum, first quartile, median, third quartile, and maximum. You can check out *Chapter 4*, *Performing Univariate Analysis in Python*, for more details about the boxplot and its components. When used on time series data, the x axis typically represents our time intervals, while the y axis represents a variable of interest. The time intervals on the x axis are typically summarized time intervals, for example, hourly summarized to daily, daily summarized to monthly, or yearly.

The boxplot shows how the mean of our variable on the y axis changes over time. It also shows how dispersed the data points are within a time interval. The axes of the boxplot need to be scaled correctly to avoid misleading results. A wrong scaling of the y axis can lead to more outliers than expected.

We will explore how to visualize time series data using the line chart and boxplot. We will use the `plot` function in `matplotlib` and the `boxplot` function in `seaborn` to do this.

Getting ready

In this recipe, we will work with one dataset: San Francisco Air Traffic Passenger Statistics data from Kaggle.

Create a folder for this chapter and create a new Python script or Jupyter Notebook file in that folder. Create a data subfolder and place the `SF_Air_Traffic_Passenger_Statistics_Transformed.csv` file in that subfolder. Alternatively, you could retrieve all the files from the GitHub repository.

> **Note**
>
> Kaggle provides the San Francisco Air Traffic Passenger Statistics data for public use data at `https://www.kaggle.com/datasets/oscarm524/san-francisco-air-traffic-passenger-statistics?select=SF_Air_Traffic_Passenger_Statistics.csv`. However, for this recipe and others in this chapter, the data was aggregated to give only the date and the total passenger count. The original and transformed data are available in the repository.

How to do it...

We will learn how to create a line plot and boxplot using the `matplotlib` and `seaborn` libraries:

1. Import the `numpy`, `pandas`, `matplotlib`, and `seaborn` libraries:

    ```
    import numpy as np
    import pandas as pd
    import matplotlib.pyplot as plt
    import seaborn as sns
    ```

2. Load the `.csv` into a dataframe using `read_csv`:

    ```
    air_traffic_data = pd.read_csv("data/SF_Air_Traffic_Passenger_
    Statistics_Transformed.csv")
    ```

3. Check the first five rows using the `head` method. Check the number of columns and rows as well as the data types:

    ```
    air_traffic_data.head()
          Date       Total Passenger Count
    0     200601         2448889
    1     200602         2223024
    2     200603         2708778
    3     200604         2773293
    4     200605         2829000

    air_traffic_data.shape
    (132, 2)

    air_traffic_data.dtypes
    Date                      int64
    Total Passenger Count     int64
    ```

4. Convert the `Date` column from the `int` data type to the `datetime` data type:

    ```
    air_traffic_data['Date']= pd.to_datetime(air_traffic_
    data['Date'], format = "%Y%m")

    air_traffic_data.dtypes
    Date                      datetime64[ns]
    Total Passenger Count         int64
    ```

5. Set Date as the index of the dataframe:

```
air_traffic_data.set_index('Date',inplace = True)
air_traffic_data.shape
(132,1)
```

6. Plot the time series data on a line plot using the plot function in matplotlib:

```
plt.figure(figsize= (18,10))


plt.plot(air_traffic_data.index, air_traffic_data['Total
Passenger Count'], color='tab:red')
plt.title("Time series analysis of San Franscisco Air
Traffic",fontsize = 20)
plt.ticklabel_format(style='plain', axis='y')
```

This results in the following output:

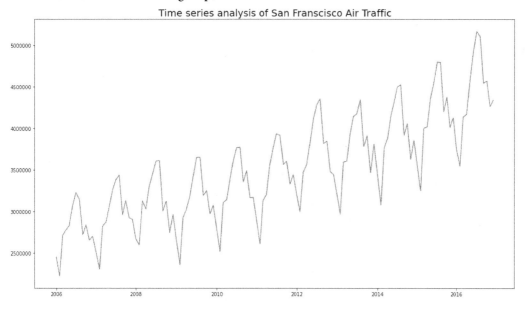

Figure 7.2: Line plot of the air traffic time series data

7. Plot the time series data on a boxplot using the boxplot function in seaborn:

```
plt.figure(figsize= (18,10))


data_subset = air_traffic_data[air_traffic_data.index < '2010-
01-01']
```

```
ax = sns.boxplot(data = data_subset , x= pd.PeriodIndex(data_
subset.index, freq='Q'),
                    y= data_subset['Total Passenger Count'])

ax.set_title("Time series analysis of San Franscisco Air
Traffic",fontsize = 20)

plt.ticklabel_format(style='plain', axis='y')
```

This results in the following output:

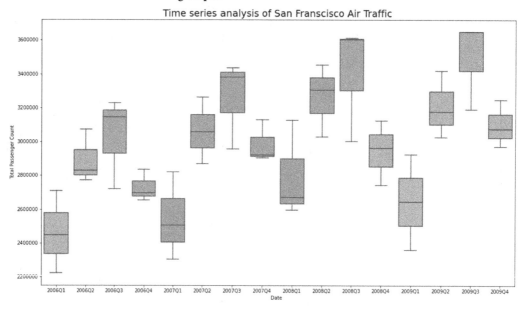

Figure 7.3: Boxplot of the air traffic time series data

Great. Now, we have created a line plot and boxplot to visualize our time series data.

How it works...

In this recipe, we use the pandas, matplotlib, and seaborn libraries. We import pandas as pd, matplotlib as plt, and seaborn as sns. In *step 2*, we load the .csv file into a dataframe using the read_csv function. In *step 3*, we examine the first five rows using the head method, check the shape (number of rows and columns) using shape, and check the data types using the dtypes attribute.

In *step 4*, we convert the date column from an integer data type to a date data type using the `to_datetime` function in `pandas`. We use the `format` parameter to specify the current format of the date. In *step 5*, we set the date column as the index of the dataframe using the `set_index` method in `pandas`. This is good practice because often series operations in Python require a date index on time series.

In *step 6*, we use the `plot` function in `matplotlib` to create a line plot of the time series data. We specify the date index column as the first argument, the passenger count column as the second argument, and the preferred line color as the third argument. We use the `ticklabel_format` function to suppress the scientific label on the *y* axis. The `matplotlib` library often uses scientific labels for continuous scales.

In *step 7*, we use the `boxplot` function in `seaborn` to create a boxplot of the time series data. First, we create a data subset by specifying a date range. This range ensures the data is easier to visualize and the boxplot is not too rowdy. In the `boxplot` function, the value of the first parameter is the dataset, the value of the second parameter is the date index column, and the value of the third parameter is the passenger count. For the value of the second parameter (date index column), we apply the `pandas` `PeriodIndex` function to split the date into quarters per year.

Spotting patterns in time series

There are four types of patterns we typically should look out for when analyzing time series data. This includes trends, seasonal variations, cyclical variations, and irregular variations. Line plots are very helpful charts for analyzing these patterns. With line plots, we can easily spot these patterns within our dataset.

When analyzing our data for trends, we try to spot a long-term increase or decrease in the values in the time series. When analyzing our data for seasonal variations, we try to identify periodic patterns that are influenced by the calendar (quarter, month, day of the week, and so on). When analyzing our data for cyclical variations, we try to spot sections where the data points rise and fall with varying magnitudes and over longer periods which aren't fixed; for example, the duration of a cycle is at least two years, and each cycle can occur over a range of years (such as every two to four years) and not a fixed period (such as every two years).

We will spot patterns in time series data using the line plot and box plot. We will use the `plot` function in `matplotlib` and the `boxplot` function in `seaborn` to do this.

Getting ready

We will work with the San Francisco Air Traffic Passenger Statistics data from Kaggle in this recipe. You can retrieve all the files from the GitHub repository.

How to do it...

We will learn how to create a line plot and boxplot using the `matplotlib` and `seaborn` libraries:

1. Import the numpy, pandas, matplotlib, and seaborn libraries:

    ```
    import numpy as np
    import pandas as pd
    import matplotlib.pyplot as plt
    import seaborn as sns
    ```

2. Load the .csv into a dataframe using `read_csv`:

    ```
    air_traffic_data = pd.read_csv("data/SF_Air_Traffic_Passenger_
    Statistics_Transformed.csv")
    ```

3. Check the first five rows using the `head` method. Check the number of columns and rows as well as the data types:

    ```
    air_traffic_data.head()
            Date        Total Passenger Count
    0       200601      2448889
    1       200602      2223024
    2       200603      2708778
    3       200604      2773293
    4       200605      2829000

    air_traffic_data.shape
    (132, 2)

    air_traffic_data.dtypes
    Date                    int64
    Total Passenger Count   int64
    ```

4. Convert the `Date` column from the `int` data type to the `datetime` data type:

    ```
    air_traffic_data['Date']= pd.to_datetime(air_traffic_
    data['Date'], format = "%Y%m")

    air_traffic_data.dtypes
    Date                    datetime64[ns]
    Total Passenger Count           int64
    ```

5. Set `Date` as the index of the dataframe:

    ```
    air_traffic_data.set_index('Date',inplace = True)
    air_traffic_data.shape
    (132,1)
    ```

6.   Plot the time series data on a line plot using the `plot` function in `matplotlib`:

```
plt.figure(figsize= (18,10))

plt.plot(air_traffic_data.index, air_traffic_data['Total
Passenger Count'], color='tab:red')
plt.title("Time series analysis of San Franscisco Air
Traffic",fontsize = 20)
plt.ticklabel_format(style='plain', axis='y')
```

This results in the following output:

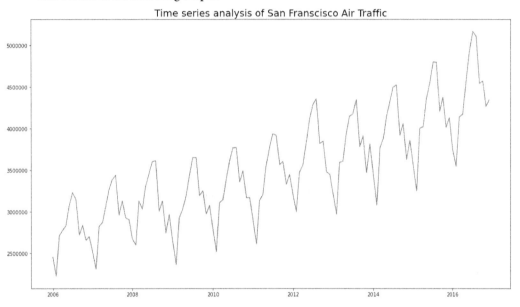

Figure 7.4: Line plot of the air traffic time series data

7.   Plot the time series data on a boxplot using the `boxplot` function in `seaborn`:

```
plt.figure(figsize= (18,10))

data_subset = air_traffic_data[air_traffic_data.index < '2010-
01-01']

ax = sns.boxplot(data = data_subset , x= pd.PeriodIndex(data_
subset.index, freq='Q'),
                y= data_subset['Total Passenger Count'])

ax.set_title("Time series analysis of San Franscisco Air
Traffic",fontsize = 20)

plt.ticklabel_format(style='plain', axis='y')
```

This results in the following output:

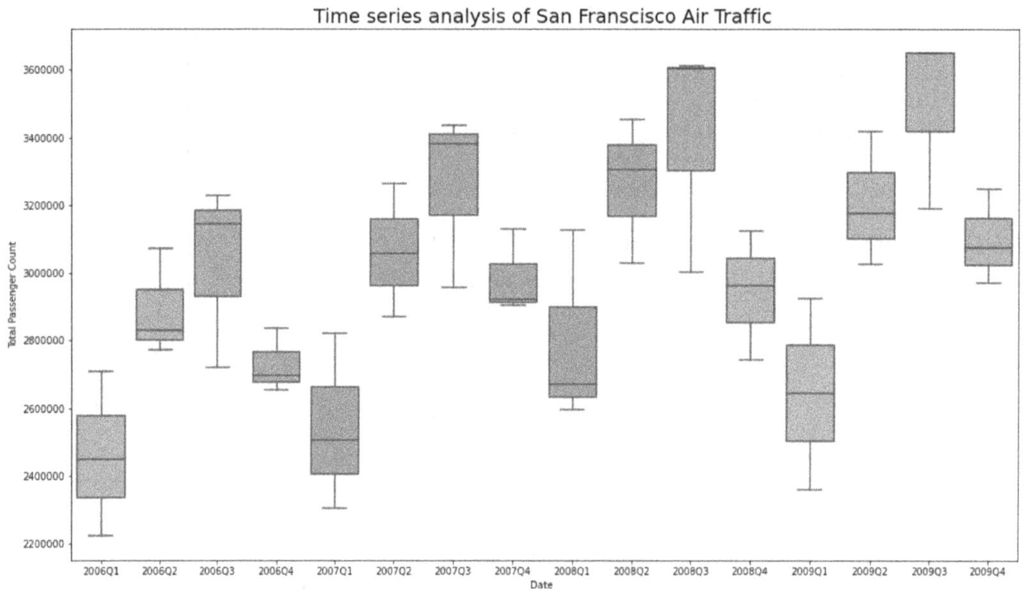

Figure 7.5: Boxplot of the air traffic time series data

Great. Now, we have created a line plot and a boxplot to visualize our time series data.

How it works...

In *step 1*, we import the relevant libraries. In *step 2*, we load the `.csv` file into a `pandas` dataframe using `read_csv`.

In *steps 3 to 5*, we inspect the dataset using the `head` method and `dtypes`, and `shape` attributes. We convert the date column from an integer data type to a date data type using the `to_datetime` function. We also set the date column as the dataframe index using the `set_index` method. This is required for many time series operations in Python.

In *step 6*, we use the `plot` function in `matplotlib` to create a line plot of the time series data. Using the date index column as the first argument, the passenger count column as the second argument, and the preferred line color as the third argument. Using the `ticklabel_format` function to suppress the scientific label on the *y* axis.

In *step 7*, we use the `boxplot` function in `seaborn` to create a boxplot of the time series data. We use a date range to create a data subset that will allow us to visualize the data better. In the `boxplot` function, the value of the first parameter is the dataset, the value of the second parameter is the date index column, and the value of the third parameter is the passenger count. For the value of the second parameter (date index column), we apply the `pandas PeriodIndex` function to split the date into quarters per year.

From the line plot, we can spot an upward trend, given the rising values of the passenger count. We can also spot seasonality by the consistent yearly fluctuations we see each year; we notice significant spikes around mid-year. This is likely to be summer. The time series data has no cyclical fluctuations. From the boxplot, we can spot fluctuations across the quarters in the year. Quarters 2 (**Q2**) and 3 (**Q3**) seem to have the highest passenger counts across the years.

Performing time series data decomposition

Decomposition is the process of splitting time series data into individual components to gain better insights into underlying patterns. In general, decomposition helps us to understand the underlying patterns in our time series better. The components are defined as follows:

- **Trend**: The long-term increase or decrease of the values in the time series

- **Seasonality**: The variations in the time series which are influenced by seasonal factors (e.g., quarter, month, week, or day)

- **Residual**: The patterns left after trend and seasonality have been accounted for. It is also considered noise (random variation in the time series)

As you may have noticed, cyclical variations covered in previous recipes do not appear as a component in decomposed time series. It is usually combined with the trend component and called trend.

When decomposing our time series, we can consider the time series as either an additive or multiplicative combination of the trend, seasonality, and residual components. The formula for each model is shown here:

Additive Model

$Y = T + S + R$

Multiplicative Model

$Y = T \times S \times R$

Y is the series, T is the trend, S is the seasonality, and R is the residual component.

The additive model is appropriate when the seasonal variation is relatively constant over time, while the multiplicative model is appropriate when the seasonal variation increases over time. In the multiplicative model, the size/magnitude of the fluctuations aren't constant but increase over time.

We will explore additive and multiplicative decomposition models using `seasonal_decompose` in `statsmodel`.

Multiplicative decomposition models can be represented as follows:

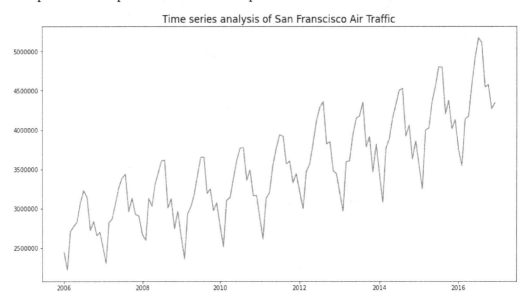

Figure 7.6: Time series for multiplicative decomposition

The following graph showcases additive decomposition:

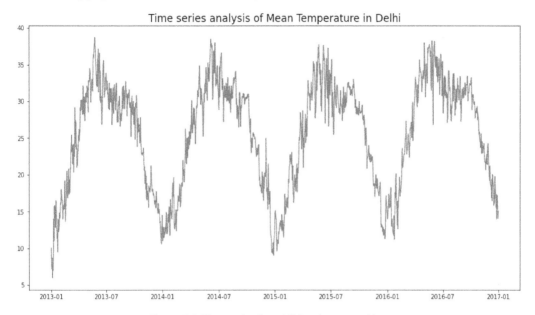

Figure 7.7: Time series for additive decomposition

Getting ready

In this recipe, we will work with two datasets: San Francisco Air Traffic Passenger Statistics data from Kaggle and Daily Climate time series data from Kaggle.

> **Note**
>
> Kaggle also provides the Daily Climate time series data for public use data at `https://www.kaggle.com/datasets/sumanthvrao/daily-climate-time-series-data`. The data is available in the repository.

How to do it...

We will learn how to decompose time series data using the `statsmodel` library:

1. Import the `numpy`, `pandas`, `matplotlib`, `seaborn`, and `statsmodel` libraries:

   ```
   import numpy as np
   import pandas as pd
   import matplotlib.pyplot as plt
   ```

```
import seaborn as sns
from statsmodels.tsa.seasonal import seasonal_decompose
```

2. Load the .csv files into a dataframe using read_csv:

```
air_traffic_data = pd.read_csv("data/SF_Air_Traffic_Passenger_
Statistics_Transformed.csv")
weather_data = pd.read_csv("data/DailyDelhiClimate.csv")
```

3. Check the first five rows using the head method. Check the number of columns and rows as well as the data types:

```
air_traffic_data.head()
      Date       Total Passenger Count
0     200601         2448889
1     200602         2223024


weather_data.head(2)


air_traffic_data.shape
(132, 2)

weather_data.shape
(1461, 2)

air_traffic_data.dtypes
Date                     int64
Total Passenger Count    int64

weather_data.dtypes
date          object
meantemp      float64
```

4. Convert the Date column from the int data type to the datetime data type:

```
air_traffic_data['Date'] = pd.to_datetime(air_traffic_
data['Date'], format = "%Y%m")


air_traffic_data.dtypes
Date                     datetime64[ns]
Total Passenger Count         int64


weather_data['date'] = pd.to_datetime(weather_data['date'],
format = "%d/%m/%Y")
```

```
weather_data.dtypes
date              datetime64[ns]
meantemp                 float64
```

5. Set `Date` as the index of the dataframe:

```
air_traffic_data.set_index('Date',inplace = True)
air_traffic_data.shape
(132,1)
weather_data.set_index('date', inplace = True)
weather_data.shape
(1461,1)
```

6. Plot the air traffic time series data on a line plot using the `plot` function in `matplotlib`:

```
plt.figure(figsize= (15,8))

plt.plot(air_traffic_data.index, air_traffic_data['Total
Passenger Count'], color='tab:red')
plt.title("Time series analysis of San Franscisco Air
Traffic",fontsize = 17)
plt.ticklabel_format(style='plain', axis='y')
```

This results in the following output:

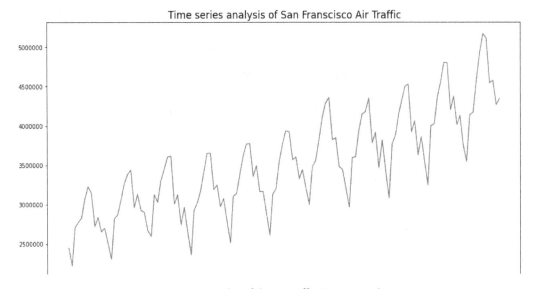

Figure 7.8: Line plot of the air traffic time series data

7. Plot the weather time series data on a line plot using the `plot` function in `matplotlib`:

```
plt.figure(figsize= (15,8))

plt.plot(weather_data.index, weather_data['meantemp'],
color='tab:blue')
plt.title("Time series analysis of Mean Temperature in
Delhi",fontsize = 17)
plt.ticklabel_format(style='plain', axis='y')
```

This results in the following output:

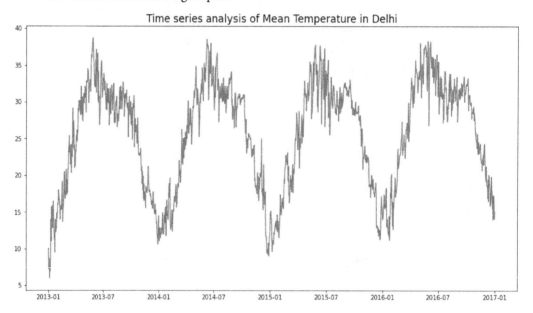

Figure 7.9: Line plot of the weather time series data

8. Decompose the air traffic time series data using a multiplicative model:

```
decomposition_multi = seasonal_decompose(air_traffic_data['Total
Passenger Count'],

    model='multiplicative', period = 12)

fig = decomposition_multi.plot()
fig.set_size_inches((10, 9))
plt.show()
```

This results in the following output:

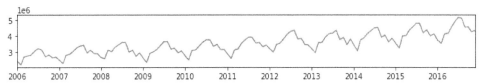

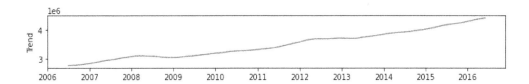

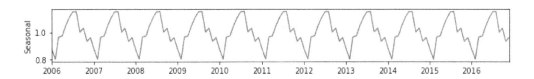

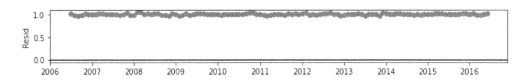

Figure 7.10: Multiplicative decomposition of the Air traffic time series data

9. Decompose the weather time series data using an additive model:

```
decomposition_add = seasonal_decompose(weather_data['meantemp'],

    model='additive',period = 365)

fig = decomposition_add.plot()
fig.set_size_inches((10, 9))
plt.show()
```

This results in the following output:

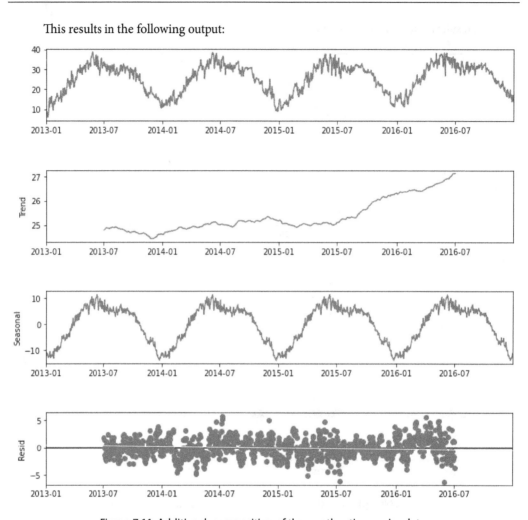

Figure 7.11: Additive decomposition of the weather time series data

Awesome. We have decomposed our time series data using the additive and multiplicative models.

How it works...

In *steps 1* to *2*, we import the relevant libraries and load the `.csv` files containing the air traffic data and the weather data into two separate `pandas` dataframes.

In *steps 3* to *5*, we inspect both datasets using the `head method`, `dtypes`, and `shape` attributes. We convert the date column in the air traffic data from an integer data type to a date data type using the `to_datetime` function. We use the same function to convert the date column in the weather data from an object data type to a date data type. We also set the date column as the dataframe index for each dataframe. We achieve this using the `set_index` method.

In *step 6*, we use the `plot` function in `matplotlib` to create a line plot of the air traffic time series data to get a sense of what the data looks like. In *step 7*, we use the same function to get a view of what the weather data looks like. In *step 8*, we use the `seasonal_decompose` class in `statsmodel` to decompose the air traffic data. A multiplicative model is used because the seasonal variation increases over time. We specify the data as the first argument, the type of model as the second argument, and the period as the third. The `period` parameter is used to indicate the number of observations in a seasonal cycle. A period of 12 was used because we are assuming that the seasonal pattern repeats every year, and since our data has a monthly frequency, there are 12 observations in a year.

In *step 9*, we use the `seasonal_decompose` class in `statsmodel` to decompose the weather data. An additive model is used because the seasonal variation is constant over time. Again, we specify the data as the first argument, the type of model as the second argument, and the period as the third. A period of 365 was used because we are assuming that the seasonal pattern repeats every year, and since our data has a daily frequency, there are 365 observations in a year.

The decomposition gives the following components in its output:

- **Original time series**: This is a line plot displaying the original time series.
- **Trend**: This displays the long-term increase/decrease of the values in the time series. In the air traffic time series data, we can spot a strong trend, which is a result of the large changes in passenger count over the 10-year period (note that the *y* axis of the trend chart is in millions). On the other hand, the weather time series data shows a weak trend over the four-year period, given the small changes to mean temperature during this time.
- **Seasonal**: This displays the variations that are influenced by seasonal factors. Both time series data show strong seasonality patterns.
- **Residual**: This displays fluctuations left after trend and seasonality have been accounted for. The air traffic time series data shows very minimal residual values, and this means that the trend and seasonality account for most of the fluctuations in the data. However, the weather time series data has higher residual values, which means the trend and seasonality do not account for a significant proportion of the fluctuations in the data.

Performing smoothing – moving average

Smoothing is usually performed as a part of time series analysis to reveal patterns and trends better. Smoothing techniques can help remove noise from our time series dataset; this reduces the impact of short-term fluctuations and reveals underlying patterns better.

Moving average is a commonly used smoothing technique that helps identify trends and patterns in time series data. It works by computing the average of a set of adjacent data points in the time series. The number of adjacent points is determined by the size of a window. The window refers to the number of periods for which we want to compute an average. This window is usually chosen based on the frequency of the data and the nature of the patterns we are trying to identify. For example, if we have daily stock price data, we can decide to compute the average price over the last 10 days, inclusive of today. In this case, 10 days is the window size. To achieve the "moving" part, we will need to slide the window along the time series to calculate the average values. In simple terms, this means on January 20, the 10-day window will span from January 11 to January 20, while on January 21, the window will span from January 12 to January 21.

Typically, we can visualize the moving average against the original time series, in order to identify any trends or patterns that are present. Short windows are often used to capture short-term fluctuations or rapid changes in the data, while long windows are used to capture longer-term trends and smoothen out short-term fluctuations.

In general, smoothing can also be used to identify outliers or sudden changes in the time series.

We will explore the moving average smoothing technique in Python. We will use the `rolling` and `mean` methods in `pandas`.

Getting ready

In this recipe, we will work with one dataset: MTN Group Limited stocks data from Yahoo Finance.

> **Note**
> Kaggle provides the MTN Group Limited stocks data for public use at `https://finance.yahoo.com/quote/MTNOY/history?p=MTNOY`. The data is available in the repository.

How to do it...

We will learn how to compute the moving average using `pandas`:

1. Import the `numpy`, `pandas`, `matplotlib`, and `seaborn` libraries:

    ```
    import numpy as np
    import pandas as pd
    import matplotlib.pyplot as plt
    import seaborn as sns
    ```

2. Load the `.csv` into a dataframe using `read_csv` and subset only the relevant columns:

```
stock_data = pd.read_csv("data/MTNOY.csv")
stock_data = stock_data[['Date','Close']]
```

3. Check the first five rows using the `head` method. Check the number of columns and rows as well as the data types:

```
stock_data.head()
        Date       Close
   2010-02-01      14.7
   2010-02-02      14.79
   2010-02-03      14.6
   2010-02-04      14.1
   2010-02-05      14.28


stock_data.shape
(1490, 2)


stock_data.dtypes
Date       object
Close      float64
```

4. Convert the `Date` column from the `int` data type to the `datetime` data type:

```
stock_data['Date']= pd.to_datetime(stock_data['Date'], format =
"%Y-%m-%d")
stock_data.dtypes
Date       datetime64[ns]
Close               float64
```

5. Set `Date` as the index of the dataframe:

```
stock_data.set_index('Date',inplace = True)
stock_data.shape
(1490,1)
```

6. Plot the stocks data on a line plot using the `plot` function in `matplotlib`:

```
plt.figure(figsize= (18,10))
plt.plot(stock_data.index, stock_data['Close'],color='tab:blue')
```

This results in the following output:

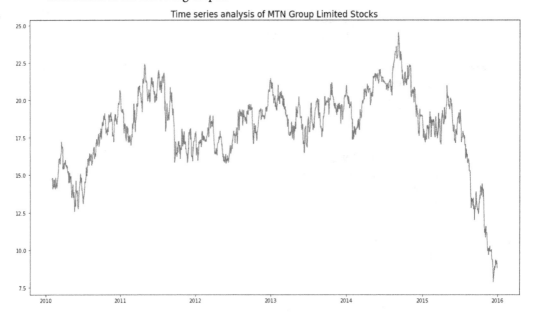

Figure 7.12: Line plot of the stocks time series data

7. Perform moving average smoothing using the `rolling` and `mean` methods in `pandas`:

```
moving_data = stock_data.rolling(window=4)
moving_average_data = moving_data.mean()
moving_average_data.head()
Date            Close
2010-02-01
2010-02-02
2010-02-03
2010-02-04        14.5475
2010-02-05        14.4425
```

8. Plot the original time series data against the moving average data on a line plot using the `plot` function in `matplotlib`:

```
plt.figure(figsize= (18,10))

stock_data_subset = stock_data[stock_data.index>= '2015-01-01']
moving_average_data_subset = moving_average_data[moving_average_
data.index>= '2015-01-01']
```

```
plt.plot(stock_data_subset.index, stock_data_
subset['Close'],color='tab:blue',alpha = 0.4)
plt.plot(moving_average_data_subset.index,moving_average_data_
subset['Close'],color='tab:red')
```

This results in the following output:

Figure 7.13: Line plot of the stocks time series data versus the moving average

9. Calculate the residuals and plot the residuals on a histogram:

```
plt.figure(figsize= (18,10))

residuals = stock_data - moving_average_data
plt.hist(residuals, bins=50)
plt.show()
```

This results in the following output:

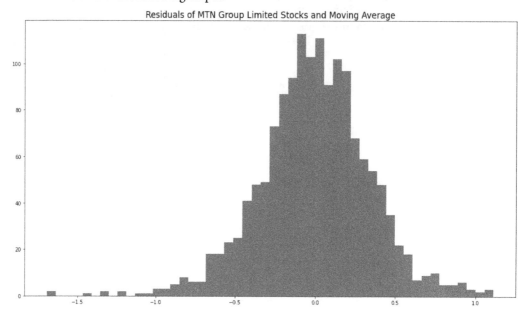

Figure 7.14: Residuals of the stocks time series data and the moving average

10. Examine the residuals with very high values:

```
percentile_90 = np.nanpercentile(abs(residuals['Close']),90)
residuals[abs(residuals['Close']) >= percentile_90]
Date           Close
2010-03-18     0.5749995000000006
2010-03-25     -1.0125002499999987
2010-05-05     -0.5699999999999985
2010-05-06     -0.6624999999999996
2010-05-20     -0.697499999999998
  ...            ...
2015-09-17     0.587499999999986
2015-10-26     -1.2249999999999996
2015-10-27     -1.3100000000000005
2015-10-28     -0.8975000000000009
2015-12-11     -0.7474999999999996
```

Great. Now, we have performed moving average smoothing on our time series data.

How it works...

In *steps 1* to *2*, we import the relevant libraries and load the `.csv` file into a dataframe using the `import` and `read_csv` commands, respectively.

In *steps 3* to *5*, we get a glimpse of the data using the `head` method, `dtypes`, and `shape` attributes. We use the `to_datetime` function to convert the date column from integer to date. We also set the date column as the dataframe index using the `set_index` method.

In *step 6*, we use the `plot` function in `matplotlib` to create a line plot of the time series data to get a sense of what the data looks like. In *step 7*, we compute the moving average. First, we use the `rolling` method in `pandas` to define our moving average window. Then we compute the average of the rolling window using the `mean` method. In *step 8*, we plot the moving average data against the original time series data. The moving average line plot smoothens out the short-term fluctuations in the data.

In *step 9*, we compute residuals and plot the output in a histogram. We compute the residuals by finding the difference between the values of the moving average and that of the original time series. We then plot the residuals on a histogram using the `hist` function in `matplotlib`. This histogram gives a sense of the shape of the distribution of residuals where the values to the extreme left and right can be considered outliers. We set a threshold of 90 percentile to review possible outliers with values beyond the 90th percentile. These values can also point us to sudden changes in the time series.

See also...

You can check out the following insightful resource:

`https://machinelearningmastery.com/moving-average-smoothing-for-time-series-forecasting-python/`

Performing smoothing – exponential smoothing

Another commonly used smoothing technique is exponential smoothing. It gives more weight to recent observations and less to the older ones. While moving average smoothing applies equal weights to past observations, exponential smoothing applies exponentially decreasing weights to observations as they get older. A major advantage it has over the moving average is the ability to capture sudden changes in the data more effectively. This is because exponential smoothing gives more weight to recent observations and less to previous ones, unlike the moving average, which applies equal weights.

Beyond time series exploratory analysis, both moving average and exponential smoothing techniques can also be used as a basis for forecasting future values in a time series.

We will explore the exponential smoothing technique in Python. We will use the `ExponentialSmoothing` module in `statsmodels`.

Getting ready

We will work with one dataset: MTN Group Limited stocks data from Yahoo Finance.

How to do it...

We will learn how to compute exponential smoothing using the `statsmodel` library:

1. Import the `numpy`, `pandas`, `matplotlib`, and `seaborn` libraries:

    ```
    import numpy as np
    import pandas as pd
    import matplotlib.pyplot as plt
    import seaborn as sns
    from statsmodels.tsa.api import ExponentialSmoothing
    ```

2. Load the `.csv` into a dataframe using `read_csv` and subset only the relevant columns:

    ```
    stock_data = pd.read_c"v("data/MTNOY."sv")
    stock_data = stock_dat'[['D't'','Cl'se']]
    ```

3. Check the first five rows using the `head` method. Check the number of columns and rows as well as the data types:

    ```
    stock_data.head()
          Date        Close
    2010-02-01       14.7
    2010-02-02       14.79
    2010-02-03       14.6
    2010-02-04       14.1
    2010-02-05       14.28


    stock_data.shape
    (1490, 2)


    stock_data.dtypes
    Date          object
    Close        float64
    ```

4. Convert the `Date` column from the `int` data type to the `datetime` data type:

    ```
    stock_da'a['D'te']= pd.to_datetime(stock_da'a['D'te'], format"=
    "%Y-%m"%d")
    stock_data.dtypes
    Date         datetime64[ns]
    Close                float64
    ```

5. Set `Date` as the index of the dataframe:

```
stock_data.set_ind'x('D'te',inplace = True)
stock_data.shape
(1490,1)
```

6. Plot the stocks data on a line plot using the `plot` function in `matplotlib`:

```
plt.figure(figsize= (18,10))
plt.plot(stock_data.index, stock_da'a['Cl'se'],col'r='tab:b'ue')
```

This results in the following output:

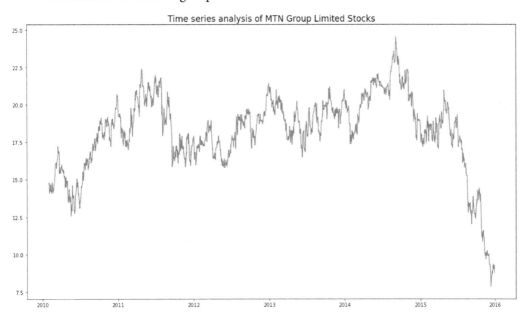

Figure 7.15: Line plot of the stocks time series data

7. Perform exponential smoothing using the `exponentialsmoothing` class in `statsmodel`:

```
exponentialmodel = ExponentialSmoothing(stock_data)
fit = exponentialmodel.fit(smoothing_level=0.7)
fitted_values = fit.fittedvalues
```

8. Plot the original time series data against the exponential smoothing data on a line plot using the `plot` function in `matplotlib`:

```
plt.figure(figsize= (18,10))

stock_data_subset = stock_data[stock_data.index'= '2015-01'01']
fitted_values_subset = fitted_values[fitted_values.index'=
'2015-01'01']

plt.plot(stock_data_subset.index, stock_data_
subs't['Cl'se'],lab'l='Original D'ta',col'r='tab:b'ue',alpha =
0.4)
plt.plot(fitted_values_subset.index,fitted_values_subset.
values,lab'l='Exponential Smooth'ng',col'r='tab:'ed')
```

This results in the following output:

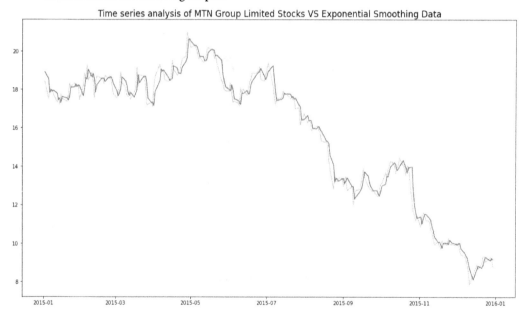

Figure 7.16: Line plot of the stocks time series data versus exponential smoothing data

9. Calculate residuals and plot the residuals on a histogram:

```
plt.figure(figsize= (18,10))

residuals = stock_data.squeeze() - fitted_values
plt.hist(residuals, bins=50)
plt.show()
```

This results in the following output:

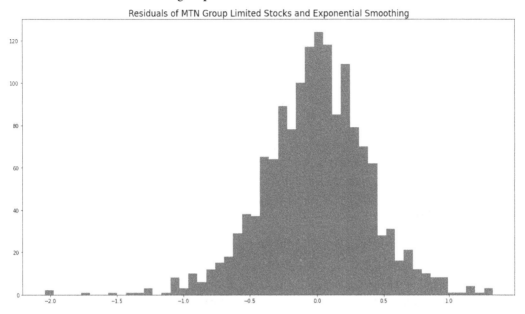

Figure 7.17: Residuals of the stocks time series data and exponential smoothing data

10. Examine the residuals with very high values:

```
percentile_90 = np.nanpercentile(abs(residuals),90)
residuals[abs(residuals) >= percentile_90]
Date
2010-03-18        0.7069166397051063
2010-03-25       -1.2724502555655164
2010-05-20       -0.714170745592277
2010-05-26        0.9421852169607021
2010-05-27        0.9426555650882111
     ...                 ...
2015-09-04       -0.663435304195815
2015-09-09       -0.933709177377624
2015-10-26       -1.7499660773721946
2015-10-27       -1.0749898232116593
2015-12-11       -0.797649483196115
```

Awesome. We just performed exponential smoothing on our time series data.

How it works...

In *steps 1* to *2*, we import the relevant libraries and load the `.csv` file into a dataframe.

In *steps 3* to *5*, we inspect the dataset using the `head` method, `dtypes`, and `shape` attributes. We convert the date column from an integer data type to a date data type using the `to_datetime` function. We also set the date column as the dataframe index using the `set_index` method. This is required for many time series operations in Python.

In *step 6*, we use the `plot` function in `matplotlib` to create a line plot of the time series data to get a sense of what the data looks like. In *step 7*, we create an exponential smoothing model using the `ExponentialSmoothing` class in `statsmodel`. Then we fit the model using the `fit` method. Fitting is the process whereby we train the model on the time series data. The `fit` method uses a `smoothing_level` parameter that controls the rate at which prior observations decay exponentially. We extract the fitted values from the `fittedvalues` attribute and save them into a `pandas` series.

In *step 8*, we plot the exponential smoothing data against the original time series data.

In *step 9*, we compute the residuals and plot the output in a histogram. We compute the residuals by finding the difference between the values of the exponential smoothing and that of the original time series. We use the `squeeze` method in `pandas` to convert the original time series dataframe into a `pandas` series since the fitted values are in a `pandas` series. We then plot the residuals on a histogram using the `hist` function in `matplotlib`. This histogram gives a sense of the shape of the distribution of residuals where the values to the extreme left and right can be considered outliers. We set a threshold of 90 percentile to review possible outliers with values beyond the 90th percentile. These values can also point us to sudden changes in the time series.

See also...

You can check out the following insightful resources:

- `https://machinelearningmastery.com/exponential-smoothing-for-time-series-forecasting-in-python/`

- The Investopedia website provides a practical example of how the moving average concept is used in investment:

  `https://www.investopedia.com/articles/active-trading/052014/how-use-moving-average-buy-stocks.asp`

Performing stationarity checks on time series data

Stationarity is an essential concept in time series. Stationary data has statistical properties such as mean, variance, and covariance, which do not change over time. Also, stationary data doesn't contain trends and seasonality; typically, time series with these patterns are called non-stationary. Checking for stationarity is important because non-stationary data can be challenging to model and predict. Overall, stationarity can help inform forecasting model selection and prediction accuracy.

To test stationarity, we can use a statistical test called the Dickey-Fuller test. Without going into technicalities, the Dickey-Fuller Test works with the following hypotheses:

- **Null hypothesis**: The time series data is non-stationary
- **Alternative hypothesis**: The time series data is stationary

The test generates a test statistic and critical values at significant levels of 1%, 5%, and 10%. We typically compare the value of the test statistic to the critical values to reach a conclusion on stationarity.

The results can be interpreted in the following way:

- If the test statistic < the critical values, we reject the null hypothesis and conclude that the time series is stationary
- If the test statistic > the critical values, we don't reject the null hypothesis and conclude that the time series is non-stationary

The Dickey-Fuller test has various variants. A common variant is the **Augmented Dickey-Fuller (ADF)** test. The ADF test performs the Dickey-Fuller test but removes autocorrelation from the series.

We will explore the ADF test in Python, using the `adfuller` module in `statsmodels`.

Getting ready

We will work with the San Francisco Air Traffic Passenger Statistics data from Kaggle in this recipe. You can retrieve all the files from the GitHub repository.

How to do it...

We will learn how to check for stationarity using the `statsmodel` library:

1. Import the numpy, `pandas`, `matplotlib`, `seaborn`, and `statsmodel` libraries:

```
import numpy as np
import pandas as pd
import matplotlib.pyplot as plt
import seaborn as sns
from statsmodels.tsa.stattools import adfuller
```

2. Load the .csv into a dataframe using read_csv:

```
air_traffic_data = pd.read_csv("data/SF_Air_Traffic_Passenger_
Statistics_Transformed.csv")
```

3. Check the first five rows using the head method. Check the number of columns and rows as well as the data types:

```
air_traffic_data.head()
      Date      Total Passenger Count
0     200601    2448889
1     200602    2223024
2     200603    2708778
3     200604    2773293
4     200605    2829000

air_traffic_data.shape
(132, 2)

air_traffic_data.dtypes
Date                     int64
Total Passenger Count    int64
```

4. Convert the Date column from the int data type to the datetime data type:

```
air_traffic_data['Date']= pd.to_datetime(air_traffic_
data['Date'], format = "%Y%m")

air_traffic_data.dtypes
Date                     datetime64[ns]
Total Passenger Count        int64
```

5. Set Date as the index of the dataframe:

```
air_traffic_data.set_index('Date',inplace = True)
air_traffic_data.shape
(132,1)
```

6. Check for stationarity through the Dickey-Fuller test using the adfuller class in statsmodel:

```
adf_result = adfuller(air_traffic_data, autolag='AIC')
adf_result
(0.7015289287377346,
 0.9898683326442054,
 13,
 118,
```

```
{'1%': -3.4870216863700767,
 '5%': -2.8863625166643136,
 '10%': -2.580009026141913},
3039.0876643475)
```

7. Format the results of the Dickey-Fuller test output:

```
print('ADF Test Statistic: %f' % adf_result[0])
print('p-value: %f' % adf_result[1])
print('Critical Values:')
print(adf_result[4])

if adf_result[0] < adf_result[4]["5%"]:
    print ("Reject Null Hypothesis - Time Series is Stationary")
else:
    print ("Failed to Reject Null Hypothesis - Time Series is
Non-Stationary")

ADF Test Statistic: 0.701529
p-value: 0.989868
Critical Values:
{'1%': -3.4870216863700767, '5%': -2.8863625166643136, '10%':
-2.580009026141913}
Failed to Reject Null Hypothesis - Time Series is Non-Stationary
```

Nice. We just checked whether our time series data was stationary.

How it works...

In *steps 1* to *2*, we import the relevant libraries and load the `.csv` file into a dataframe. In *step 3*, we examine the first five rows using the `head` method, check the shape (number of rows and columns) using `shape`, and check the data types using the `dtypes` attributes.

In *step 4*, we convert the Date data type from an integer data type to a date data type using the `to_datetime` function in `pandas`. We use the `format` parameter to specify the current format of the date. In *step 5*, we set the date column as the index of the dataframe using the `set_index` method in `pandas`.

In *step 6*, we use the `adfuller` class to perform the ADF test. We indicate the data as the first argument and use the `autolag` parameter to indicate the optimal number of lags. When set to `'AIC'`, the lag order that minimizes the **Akaike Information Criterion (AIC)** is selected. Lags are explained in the *Using correlation plots to visualize time series data* recipe. AIC is a statistical measure commonly used in model selection, where the goal is to choose the model that best balances goodness of fit with model complexity. In *step 7*, we format the output of the `adfuller` class to make it easier to understand. Since the test statistic is greater than the critical value at 5%, we fail to reject the null hypothesis and conclude the time series is non-stationary.

See also...

Here are a couple of insightful resources on stationarity:

- https://www.kdnuggets.com/2019/08/stationarity-time-series-data.html
- https://analyticsindiamag.com/complete-guide-to-dickey-fuller-test-in-time-series-analysis/

Differencing time series data

When time series data is not stationary, we can use a technique called differencing to make it stationary. Differencing involves subtracting the current values from the preceding values to remove the trends or seasonality present in the time series. First-order differencing happens when we subtract each value from the preceding value by one time period. Differencing can be done several times, and this is known as higher-order differencing. This helps us remove higher levels of trend or seasonality. However, the downside of differencing too many times is that we may lose vital information from the original time series data.

We will explore the differencing technique in Python. We will use the `diff` method in `pandas`.

Getting ready

We will work with the San Francisco Air Traffic Passenger Statistics data from Kaggle in this recipe. You can retrieve all the files from the GitHub repository.

How to do it...

We will learn how to implement differencing using the pandas libraries:

1. Import the numpy, pandas, matplotlib, and seaborn libraries:

    ```
    import numpy as np
    import pandas as pd
    import matplotlib.pyplot as plt
    import seaborn as sns
    from statsmodels.tsa.stattools import adfuller
    ```

2. Load the .csv into a dataframe using read_csv:

    ```
    air_traffic_data = pd.read_csv("data/SF_Air_Traffic_Passenger_
    Statistics_Transformed.csv")
    ```

3. Check the first five rows using the head method. Check the number of columns and rows as well as the data types:

    ```
    air_traffic_data.head()
          Date      Total Passenger Count
    0     200601        2448889
    1     200602        2223024
    2     200603        2708778
    3     200604        2773293
    4     200605        2829000

    air_traffic_data.shape
    (132, 2)

    air_traffic_data.dtypes
    Date                     int64
    Total Passenger Count    int64
    ```

4. Convert the Date column from the int data type to the datetime data type:

    ```
    air_traffic_data['Date']= pd.to_datetime(air_traffic_
    data['Date'], format = "%Y%m")

    air_traffic_data.dtypes
    Date                     datetime64[ns]
    Total Passenger Count             int64
    ```

5. Set `Date` as the index of the dataframe:

```
air_traffic_data.set_index('Date',inplace = True)
air_traffic_data.shape
(132,1)
```

6. Make the data stationary using the `diff` method in `pandas`:

```
air_traffic_data['Difference'] = air_traffic_data['Total
Passenger Count'].diff(periods=1)
air_traffic_data = air_traffic_data.dropna()
```

7. Plot the stationary time series against the original data using the `plot` function in `matplotlib`:

```
plt.figure(figsize= (18,10))

plt.plot(air_traffic_data['Total Passenger Count'], label='Total
Passenger Count')
plt.plot(air_traffic_data['Difference'], label='First-order
difference', color='tab:red')
plt.title('Total Passenger Count VS Passenger Count with first-
order difference', size=15)
plt.legend()
```

This results in the following output:

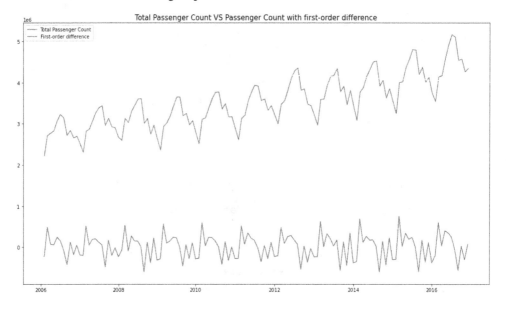

Figure 7.18: Line plot of the air traffic time series data against its first-order difference

8. Check for stationarity through the Dickey-Fuller test using the `adfuller` class in `statsmodel`:

```
adf_result = adfuller(air_traffic_data['Difference'],
autolag='AIC')
adf_result
(-3.127564931502872,
 0.024579379461460212,
 12,
 118,
 {'1%': -3.4870216863700767,
  '5%': -2.8863625166643136,
  '10%': -2.580009026141913},
 3013.039106024442)
```

9. Format the results of the Dickey-Fuller test output:

```
print('ADF Test Statistic: %f' % adf_result[0])
print('p-value: %f' % adf_result[1])
print('Critical Values:')
print(adf_result[4])

if adf_result[0] < adf_result[4]["5%"]:
    print ("Reject Null Hypothesis - Time Series is Stationary")
else:
    print ("Failed to Reject Null Hypothesis - Time Series is
Non-Stationary")

ADF Test Statistic: -3.127565
p-value: 0.024579
Critical Values:
{'1%': -3.4870216863700767, '5%': -2.8863625166643136, '10%':
-2.580009026141913}
Reject Null Hypothesis - Time Series is Stationary
```

Well done. We just performed differencing on our time series data to make it stationary.

How it works...

In *steps 1* to *2*, we import the relevant libraries and load the `.csv` file into a dataframe.

In *steps 3* to *5*, we get a glimpse of the dataset using the `head method`, `dtypes`, and `shape` attributes. We change the `Date` column data type to a date from integer using the `to_datetime` function. We also set the `Date` column as the dataframe index using the `set_index` method.

In *step 6*, we perform differencing on our time series to make it stationary. We use the `diff` method in `pandas` to achieve this. In the method, we specify the value of the `period` parameter as 1, meaning the difference between the current value and the prior value. We then drop the **Not a Number (NaN)** values as this is a requirement for the ADF test.

In *step 7*, we plot the original time series against the differenced version using the `plot` function in `pandas`. In *step 8*, we use the `adfuller` class to perform the ADF test. We specify the data as the first argument and use the `autolag` parameter to indicate the optimal number of lags. When set to `'AIC'`, the lag order that minimizes the AIC is selected. Lags are explained in the *Using correlation plots to visualize time series data* recipe. In *step 9*, we format the output of the `adfuller` class to make it easier to understand. Since the test statistic is less than the critical value at 5%, we reject the null hypothesis and conclude that the time series is stationary.

Using correlation plots to visualize time series data

Autocorrelation refers to the degree of similarity between current values in a time series and historical values of the same time series data. The current and historical values are typically separated by a time interval called a lag. The lag between a current value and a previous one is a lag of one.

By checking for autocorrelation, we are interested in knowing whether the values that are separated by a time interval have a strong positive or negative correlation. This correlation usually indicates that historical values influence current values. It helps to identify the presence of trends and other patterns in the time series data.

Partial autocorrelation is very similar to autocorrelation in that we also check for the similarity between current values in a time series and historical values of the same time series data. However, in partial autocorrelation, we remove the effect of other variables. In simple terms, we exclude the intermediary effects and focus only on the direct effects of the historical value.

For example, we may want to know the direct relationship between the number of passengers in the current period and the number four months ago. We may just want to focus on the value in the current period and the values four months ago without considering the months between. However, the values between both periods form a chain and are sometimes related. This is because the number of passengers four months ago may be related to the number three months ago, which in turn, may be related to the number two months ago all the way down to the current period. Partial autocorrelation excludes the effects of the periods in between, that is, month two and month three. It focuses on the values in the current period against the values four months ago.

We will use the `acf` and `pacf` modules in `statsmodels` to perform autocorrelation and partial autocorrelation, respectively. We will also use the `plot_acf` and `plot_pacf` modules in `statsmodels` to create the autocorrelation and partial autocorrelation plots.

Getting ready

We will work with the San Francisco Air Traffic Passenger Statistics data from Kaggle in this recipe. You can retrieve all the files from the GitHub repository.

How to do it...

We will learn how to compute autocorrelation and partial autocorrelation using the `statsmodel` library:

1. Import the `numpy`, `pandas`, `matplotlib`, and `seaborn` libraries:

    ```
    import numpy as np
    import pandas as pd
    import matplotlib.pyplot as plt
    import seaborn as sns
    from statsmodels.tsa.stattools import acf, pacf
    from statsmodels.graphics.tsaplots import plot_acf, plot_pacf
    ```

2. Load the `.csv` into a dataframe using `read_csv`:

    ```
    air_traffic_data = pd.read_csv("data/SF_Air_Traffic_Passenger_
    Statistics_Transformed.csv")
    ```

3. Check the first five rows using the `head` method. Check the number of columns and rows as well as the data types:

    ```
    air_traffic_data.head()
        Date        Total Passenger Count
    0   200601      2448889
    1   200602      2223024
    2   200603      2708778
    3   200604      2773293
    4   200605      2829000

    air_traffic_data.shape
    (132, 2)

    air_traffic_data.dtypes
    Date                     int64
    Total Passenger Count    int64
    ```

4. Convert the `Date` column from the `int` data type to `datetime` data type:

```
air_traffic_data['Date']= pd.to_datetime(air_traffic_
data['Date'], format = "%Y%m")

air_traffic_data.dtypes
Date                            datetime64 [ns]
Total Passenger Count                  int64
```

5. Set `Date` as the index of the dataframe:

```
air_traffic_data.set_index('Date',inplace = True)
air_traffic_data.shape
(132,1)
```

6. Compute autocorrelation using the `acf` class in `statsmodels`:

```
acf_values = acf(air_traffic_data[['Total Passenger
Count']],nlags = 24)
for i in range(0,25):
    print("Lag " ,i, " " , np.round(acf_values[i],2))

Lag  0    1.0
Lag  1    0.86
Lag  2    0.76
Lag  3    0.61
...       ...
Lag  11   0.67
Lag  12   0.75
Lag  13   0.63
Lag  14   0.55
...       ...
Lag  22   0.41
Lag  23   0.48
Lag  24   0.56
```

7. Plot the autocorrelation plot using the `plot_acf` class in `statsmodels`:

```
fig= plot_acf(air_traffic_data['Total Passenger Count'],lags =
24)
fig.set_size_inches((10, 9))
plt.show()
```

This results in the following output:

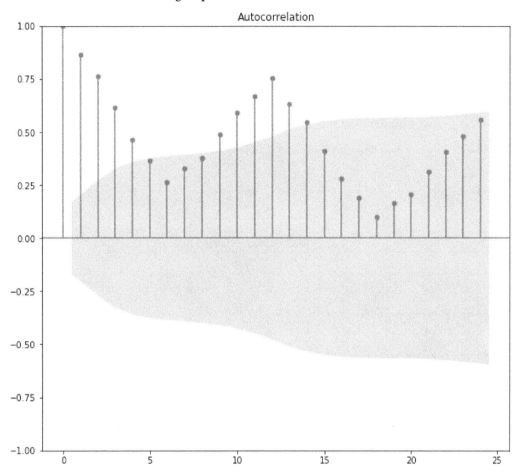

Figure 7.19: Autocorrelation of the air traffic time series data

8. Compute partial autocorrelation using the `pacf` class in `statsmodels`:

```
pacf_values = pacf(air_traffic_data['Total Passenger
Count'],nlags=24,method="ols")
for i in range(0,25):
    print("Lag " ,i, " " , np.round(pacf_values[i],2))
Lag   0    1.0
Lag   1    0.88
Lag   2    0.05
Lag   3    -0.24
...        ...
Lag   11   0.44
```

```
Lag   12    0.74
Lag   13   -0.63
Lag   14   -0.19
...        ...
Lag   22    0.24
Lag   23    0.09
Lag   24    0.12
```

9. Plot the autocorrelation plot using the `plot_pacf` class in `statsmodels`:

```
fig = plot_pacf(air_traffic_data['Total Passenger Count'],method
= 'ols', lags=24)
fig.set_size_inches((10, 9))
plt.show()
```

This results in the following output:

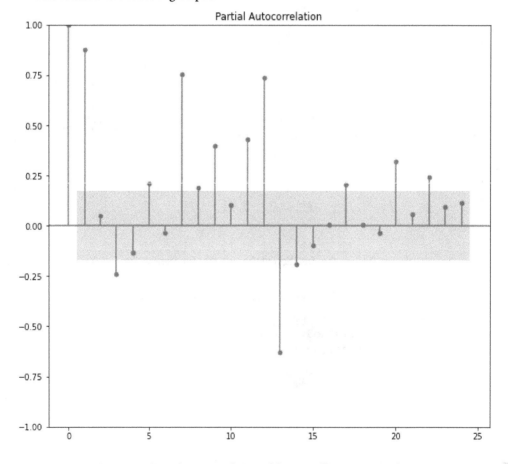

Figure 7.20: Partial Autocorrelation of the air traffic time series data

Great. Now we have created autocorrelation and partial autocorrelation plots to visualize patterns in our time series data.

How it works...

In *steps 1* to *2*, we import the relevant libraries and load the `.csv` file into a dataframe.

In *steps 3* to *5*, we inspect the dataset using the `head`, `dtypes`, and `shape` methods. We convert the `Date` column from an integer data type to a date data type using the `to_datetime` function. We also set the `Date` column as the dataframe index using the `set_index` method.

In *step 6*, we use the `acf` class to compute the autocorrelation. We specify the data as the first argument and use the `nlags` parameter to specify the number of periods. We use `24` in order to compute the autocorrelation for 24 periods, meaning 24 months. In *step 7*, we use the `plot_acf` class to plot the autocorrelation. The blue-shaded sections in the chart represent confidence intervals. The values inside the intervals are not regarded as statistically significant at the 5% level, while the values outside are statistically significant. In the chart, we notice that the shorter/closer lags have a high positive correlation; this is because, in our time series data, the values closer in time have similar values. This means that previous values within a short lag have a strong influence on current values. Also, we notice that the lags at the point of the time series seasonal frequency have a high positive correlation with each other. For example, the lags around lag 12 have a high positive correlation with lag 0. This means the value 12 months prior (lag 12) has a strong influence on the value at lag 0. This points to the seasonality that exists in the time series.

In *step 8*, we use the `pacf` class to compute the partial autocorrelation. We specify the data as the first argument and use the `nlags` parameter to specify the number of periods and the `method` parameter to specify the method (in this case, ordinary least squares). In *step 9*, we use the `plot_pacf` class to plot the partial autocorrelation. Based on the output, we see a high correlation value for lag 12. This means that the value 12 periods prior (lag 12) has a strong influence on the current value (lag 0). However, we can see that the value at lag 24 has decreased significantly. This means the value 24 months prior (lag 24) doesn't have a very strong influence on the current value (lag 0).

See also...

You can check out the following useful resources on autocorrelation and partial correlation:

- `https://towardsdatascience.com/a-step-by-step-guide-to-calculating-autocorrelation-and-partial-autocorrelation-8c4342b784e8`

- *Forecasting: Principles and Practice* by George Athanasopoulos and Rob J. Hyndman

8

Analysing Text Data in Python

Very often, we will need to perform exploratory data analysis on text data. The amount of digital data created in the world today has increased significantly, and text data forms a substantial proportion of this digital data. Some common examples we see every day include emails, social media posts, and text messages. Text data is classified as unstructured data because it usually doesn't appear in rows and columns.

In previous chapters, we focused on exploratory data analysis techniques for structured data (i.e., data that appears in rows and columns). However, in this chapter, we will focus on exploratory data analysis techniques for a very common type of unstructured data – text data.

In this chapter, we will discuss common techniques to prepare and analyze text data. The chapter includes the following:

- Preparing text data
- Removing stop words
- Analyzing **part of speech** (**POS**)
- Performing stemming and lemmatization
- Analysing Ngrams
- Creating word clouds
- Checking term frequency
- Checking sentiments
- Performing topic modeling
- Choosing an optimal number of topics

Technical requirements

We will leverage the `pandas`, `matplotlib`, `seaborn`, `nltk`, `scikit-learn`, `gensim`, and `pyLDAvis` libraries in Python for this chapter. The code and notebooks for this chapter are available on GitHub at `https://github.com/PacktPublishing/Exploratory-Data-Analysis-with-Python-Cookbook`.

Across many of the recipes, we will be using `nltk`, which is a widely used library for several text analysis tasks, such as text cleaning, stemming, lemmatization, and sentiment analysis. It has a suite of text-processing modules that perform these tasks, and it is very easy to use. To install `nltk`, we will use the code here:

```
pip install nltk
```

To perform various tasks, `nltk` also requires some resources to be downloaded. The following code helps to achieve this:

```
nltk.download('punkt')
nltk.download('stopwords')
nltk.download('wordnet')
nltk.download('averaged_perceptron_tagger')
nltk.download('vader_lexicon')
nltk.download('omw-1.4')
```

Preparing text data

Text data must go through some processing before it can be effectively analyzed. This is because text data is often messy. It could contain irrelevant information and sometimes isn't in a structure that can be easily analyzed. Some common steps for preparing text data include the following:

- **Expanding contractions**: A contraction is a shortened version of a word. It is formed by removing some letters and replacing them with apostrophes. Examples include *don't* instead of *do not*, and *would've* instead of *would have*. Usually, when preparing text data, all contractions should be expanded to their original form.

- **Removing punctuations**: Punctuations are useful for separating sentences, clauses, or phrases. However, they are mostly not needed for text analysis because they do not convey any significant meaning.

- **Converting to lowercase**: Text data is typically a combination of capital and lowercase letters. However, this needs to be standardized for easy analysis. Hence, all letters need to be converted to their lowercase form before analysis.

- **Performing tokenization**: Tokenization involves breaking up a text into smaller chunks, such as words or phrases. Breaking down text into smaller units helps us analyze it more effectively.

- **Removing stop words**: Stop words are words that do not add significant meaning to the text. Examples include *the*, *and*, and *a*. This is covered in more detail in the next recipe.

We will explore how to prepare text data for further analysis using the `word_tokenize` function in `nltk`, the `lower` method in Python, the `punctuation` constant in `string`, and the `fix` method in the `contractions` library.

> **Note**
>
> A **constant** is a pre-defined string value that is provided by a library.

Getting ready

We will work with one dataset in this chapter – the Sentiment Analysis of Restaurant Review dataset from Kaggle. Within some of the initial recipes, we will perform certain preprocessing steps and export the output to a `.csv` file. We will be working with these preprocessed files in most of the later recipes to avoid duplication of certain preprocessing steps.

Create a folder for this chapter and then a new Python script or Jupyter Notebook file in that folder. Create a `data` subfolder, and then place the `a1_RestaurantReviews_HistoricDump.tsv`, `cleaned_reviews_data.csv`, `cleaned_reviews_lemmatized_data.csv`, and `cleaned_reviews_no_stopwords_data.csv` files in that subfolder. Alternatively, you can retrieve all the files from the GitHub repository.

Along with the `nltk` library, we will require the `contractions` library too.

```
pip install contractions
```

> **Note**
>
> Kaggle provides the Sentiment Analysis of Restaurant Review dataset for public use at `https://www.kaggle.com/datasets/abhijeetkumar128/sentiment-analysis-of-restaurant-review`. This dataset comes as a **tab-delimited dataset** (TSV).

For this recipe, we will work with the original data in TSV. However, for other recipes, we will work with preprocessed versions of the original data, which are provided as `.csv` files.

How to do it...

We will learn how to prepare our text data for analysis using libraries such as `nltk`, `string`, `contractions`, and `pandas`:

1. Import the relevant libraries:

    ```
    import pandas as pd
    import string
    import nltk
    from nltk.corpus import stopwords
    from nltk.tokenize import word_tokenize
    import contractions
    ```

2. Load the `.tsv` into a dataframe using `read_csv`, and indicate the tab separator in the `sep` parameter:

    ```
    reviews_data = pd.read_csv("data/a1_RestaurantReviews_
    HistoricDump.tsv", sep='\t')
    ```

3. Check the first five rows using the `head` method:

    ```
    reviews_data.head()
    ```

    This results in the following output:

	Review	Liked
0	Wow... Loved this place.	1
1	Crust is not good.	0
2	Not tasty and the texture was just nasty.	0
3	Stopped by during the late May bank holiday of...	1
4	The selection on the menu was great and so wer...	1

    Figure 8.1: First 5 rows of the Reviews data

    Check the number of columns and rows:

    ```
    reviews_data.shape
    (900, 2)
    ```

4. Expand contractions within the reviews data using a `lambda` function, along with the `fix` method within the `contractions` library:

    ```
    reviews_data['no_contractions'] = reviews_data['Review'].
    apply(lambda x: [contractions.fix(word) for word in x.split()])
    reviews_data.head(7)
    ```

This results in the following output:

	Review	Liked	no_contractions
0	Wow... Loved this place.	1	[Wow..., Loved, this, place.]
1	Crust is not good.	0	[Crust, is, not, good.]
2	Not tasty and the texture was just nasty.	0	[Not, tasty, and, the, texture, was, just, nas...
3	Stopped by during the late May bank holiday of...	1	[Stopped, by, during, the, late, May, bank, ho...
4	The selection on the menu was great and so wer...	1	[The, selection, on, the, menu, was, great, an...
5	Now I am getting angry and I want my damn pho.	0	[Now, I, am, getting, angry, and, I, want, my,...
6	Honeslty it didn't taste THAT fresh.)	0	[Honeslty, it, did not, taste, THAT, fresh.)]

Figure 8.2: First 7 rows of the Reviews data without contractions

5. Convert the values in the expanded contractions column from a list to a string:

```
reviews_data['reviews_no_contractions'] = [' '.join(l) for l in
reviews_data['no_contractions']]
reviews_data.head(7)
```

This results in the following output:

	Review	Liked	no_contractions	reviews_no_contractions
0	Wow... Loved this place.	1	[Wow..., Loved, this, place.]	Wow... Loved this place.
1	Crust is not good.	0	[Crust, is, not, good.]	Crust is not good.
2	Not tasty and the texture was just nasty.	0	[Not, tasty, and, the, texture, was, just, nas...	Not tasty and the texture was just nasty.
3	Stopped by during the late May bank holiday of...	1	[Stopped, by, during, the, late, May, bank, ho...	Stopped by during the late May bank holiday of...
4	The selection on the menu was great and so wer...	1	[The, selection, on, the, menu, was, great, an...	The selection on the menu was great and so wer...
5	Now I am getting angry and I want my damn pho.	0	[Now, I, am, getting, angry, and, I, want, my,...	Now I am getting angry and I want my damn pho
6	Honeslty it didn't taste THAT fresh.)	0	[Honeslty, it, did not, taste, THAT, fresh.)]	Honeslty it did not taste THAT fresh.)

Figure 8.3: First 7 rows of the Reviews data without contractions

6. Tokenize the reviews data using the `word_tokenize` function in the `nltk` library:

```
reviews_data['reviews_tokenized'] = reviews_data['reviews_no_
contractions'].apply(word_tokenize)
reviews_data.head()
```

This results in the following output:

	Review	Liked	no_contractions	reviews_no_contractions	reviews_tokenized
0	Wow... Loved this place.	1	[Wow..., Loved, this, place.]	Wow... Loved this place.	[Wow, ..., Loved, this, place, .]
1	Crust is not good.	0	[Crust, is, not, good.]	Crust is not good.	[Crust, is, not, good, .]
2	Not tasty and the texture was just nasty.	0	[Not, tasty, and, the, texture, was, just, nas...	Not tasty and the texture was just nasty.	[Not, tasty, and, the, texture, was, just, nas...
3	Stopped by during the late May bank holiday of...	1	[Stopped, by, during, the, late, May, bank, ho...	Stopped by during the late May bank holiday of...	[Stopped, by, during, the, late, May, bank, ho...
4	The selection on the menu was great and so wer...	1	[The, selection, on, the, menu, was, great, an...	The selection on the menu was great and so wer...	[The, selection, on, the, menu, was, great, an...

Figure 8.4: First 5 rows of the tokenized Reviews data

7. Convert the reviews data to lowercase using the `lower` method in Python:

```
reviews_data['reviews_lower'] = reviews_data['reviews_
tokenized'].apply(lambda x: [word.lower() for word in x])
reviews_data.head()
```

This results in the following output:

	Review	Liked	no_contractions	reviews_no_contractions	reviews_tokenized	reviews_lower
0	Wow... Loved this place.	1	[Wow..., Loved, this, place.]	Wow... Loved this place.	[Wow, ..., Loved, this, place,]	[wow, ..., loved, this, place,]
1	Crust is not good.	0	[Crust, is, not, good.]	Crust is not good.	[Crust, is, not, good,]	[crust, is, not, good,]
2	Not tasty and the texture was just nasty.	0	[Not, tasty, and, the, texture, was, just, nas...	Not tasty and the texture was just nasty.	[Not, tasty, and, the, texture, was, just, nas...	[not, tasty, and, the, texture, was, just, nas...
3	Stopped by during the late May bank holiday of...	1	[Stopped, by, during, the, late, May, bank, ho...	Stopped by during the late May bank holiday of...	[Stopped, by, during, the, late, May, bank, ho...	[stopped, by, during, the, late, may, bank, ho...
4	The selection on the menu was great and so wer...	1	[The, selection, on, the, menu, was, great, an...	The selection on the menu was great and so wer...	[The, selection, on, the, menu, was, great, an...	[the, selection, on, the, menu, was, great, an...

Figure 8.5: First 5 rows of the lower case Reviews data

8. Remove punctuations from the reviews data using the `punctuations` constant within the `string` library:

```
punctuations = string.punctuation
reviews_data['reviews_no_punctuation'] = reviews_data['reviews_
lower'].apply(lambda x: [word for word in x if word not in
punctuations])
reviews_data.head()
```

This results in the following output:

	Review	Liked	no_contractions	reviews_no_contractions	reviews_tokenized	reviews_lower	reviews_no_punctuation
0	Wow... Loved this place.	1	[Wow..., Loved, this, place.]	Wow... Loved this place.	[Wow, ..., Loved, this, place, .]	[wow, ..., loved, this, place, .]	[wow, ..., loved, this, place]
1	Crust is not good.	0	[Crust, is, not, good.]	Crust is not good.	[Crust, is, not, good, .]	[crust, is, not, good, .]	[crust, is, not, good]
2	Not tasty and the texture was just nasty.	0	[Not, tasty, and, the, texture, was, just, nas...	Not tasty and the texture was just nasty.	[Not, tasty, and, the, texture, was, just, nas...	[not, tasty, and, the, texture, was, just, nas...	[not, tasty, and, the, texture, was, just, nasty]
3	Stopped by during the late May bank holiday of...	1	[Stopped, by, during, the, late, May, bank, ho...	Stopped by during the late May bank holiday of...	[Stopped, by, during, the, late, May, bank, ho...	[stopped, by, during, the, late, may, bank, ho...	[stopped, by, during, the, late, may, bank, ho...
4	The selection on the menu was great and so wer...	1	[The, selection, on, the, menu, was, great, an...	The selection on the menu was great and so wer...	[The, selection, on, the, menu, was, great, an...	[the, selection, on, the, menu, was, great, an...	[the, selection, on, the, menu, was, great, an...

Figure 8.6: First 5 rows of the Reviews data without punctuations

9. Convert the values in the expanded no_punctuations column from a list to a string:

```
reviews_data['reviews_cleaned'] = [' '.join(l) for l in reviews_
data['reviews_no_punctuation']]
reviews_data.head()
```

This results in the following output:

	Review	Liked	no_contractions	reviews_no_contractions	reviews_tokenized	reviews_lower	reviews_no_punctuation	reviews_cleaned
0	Wow... Loved this place.	1	[Wow..., Loved, this, place.]	Wow... Loved this place.	[Wow, ..., Loved, this, place, .]	[wow, ..., loved, this, place, .]	[wow, ..., loved, this, place]	wow ... loved this place
1	Crust is not good.	0	[Crust, is, not, good.]	Crust is not good.	[Crust, is, not, good, .]	[crust, is, not, good, .]	[crust, is, not, good]	crust is not good
2	Not tasty and the texture was just nasty.	0	[Not, tasty, and, the, texture, was, just, nas...	Not tasty and the texture was just nasty.	[Not, tasty, and, the, texture, was, just, nas...	[not, tasty, and, the, texture, was, just, nas...	[not, tasty, and, the, texture, was, just, nasty]	not tasty and the texture was just nasty
3	Stopped by during the late May bank holiday of...	1	[Stopped, by, during, the, late, May, bank, ho...	Stopped by during the late May bank holiday of...	[Stopped, by, during, the, late, May, bank, ho...	[stopped, by, during, the, late, may, bank, ho...	[stopped, by, during, the, late, may, bank, ho...	stopped by during the late may bank holiday of...
4	The selection on the menu was great and so wer...	1	[The, selection, on, the, menu, was, great, an...	The selection on the menu was great and so wer...	[The, selection, on, the, menu, was, great, an...	[the, selection, on, the, menu, was, great, an...	[the, selection, on, the, menu, was, great, an...	the selection on the menu was great and so wer...

Figure 8.7: First 5 rows of the cleaned Reviews data

Great! We have now prepared our text data for analysis.

How it works...

In this recipe, we use the `pandas`, `string`, `contractions`, and `nltk` libraries. In *step 2*, we load the reviews TSV file into a dataframe using the `read_csv` function and add the separator parameter to indicate the separator since the file is tab-delimited. In *step 3*, we examine the first five rows using the `head` method, and we check the shape (the number of rows and columns) using the `shape` attribute.

In *step 4*, we expand contractions within the reviews data using a lambda function along with the `apply` method in `pandas`. We pass the `fix` method within the `contractions` library into the lambda function; this takes every word in each cell of the dataframe and expands them wherever contractions exist. The output is a list of words within each cell of the dataframe. Note that in the text on row 7, the contracted word *didn't* has been expanded to *did not*. In *step 5*, we convert the values in the cells of the expanded contractions column from a list to a string. This is important because the list can sometimes result in unexpected outcomes when we perform additional preprocessing tasks on the column, especially if it involves `for` loops. We will use the `join` method in Python within a list comprehension to achieve this.

In *step 6*, we tokenize the new contractions column using the `word_tokenize` function in the `nltk` library. This breaks the sentences into individual words. In *step 7*, we convert the reviews data to lowercase using the `lower` method in Python. In *step 8*, we remove punctuations from the reviews data using the `punctuations` constant within the `string` library. In *step 9*, we convert the values in the cells of the punctuation column from a list to a string.

There's more...

Apart from NLTK, another powerful library for text processing and other text analysis tasks is the `spaCy` library. `spaCy` is a more modern library that can be used for various text analysis tasks, such as data preprocessing, part-of-speech tagging, and sentiment analysis. It is designed to be fast and efficient. It is more suitable for text analysis tasks that require high performance.

See also...

You can check out the following useful resource: *Natural Language Processing with Python – Analyzing Text with the Natural Language Toolkit.* `https://www.nltk.org/book/`

Dealing with stop words

Stop words are words that occur very frequently in a language. They generally do not add significant meaning to the text. Some common stop words include pronouns, prepositions, conjunctions, and articles. In the English language, examples of stop words include *a, an, the, and, is, was, of, for,* and *not.* This list may vary based on the language or context.

Before analyzing text, we should remove stop words so that we can focus on more relevant words in the text. Stop words typically do not have significant information and can cause noise within our dataset. Therefore, removing them helps us find insights easily and focus on what is most relevant.

However, the removal of stop words is highly dependent on the goal of our analysis and the type of task we perform. For example, the outcome of a sentiment analysis task can be misleading due to the removal of key stop words. This is highlighted here:

- **Sample sentence**: The food was not great.

- **Sample sentence without stop words**: Food great.

The first sentence is a negative review while the second sentence without the stop words is a positive review. Extra care must be taken before stop words are removed because sometimes when removing stop words, we can change the meaning of a sentence.

The removal of stop words is one of the steps we take in text preprocessing, a text preparation process that helps us analyze text more efficiently and accurately.

We will explore how to remove stop words using the `stopwords` module in `nltk`.

Getting ready

We will work with the *Sentiment Analysis of Restaurant Review* dataset for this recipe. However, we will work with the preprocessed version that was exported from the *Preparing text data* recipe. This version cleaned out punctuations, contractions, uppercase letters, and so on. You can retrieve all the files from the GitHub repository.

How to do it...

We will learn how to deal with stop words in text data using libraries such as `nltk`, `seaborn`, `matplotlib`, and `pandas`:

1. Import the relevant libraries:

    ```
    import pandas as pd
    import matplotlib.pyplot as plt
    import seaborn as sns
    import nltk
    from nltk.corpus import stopwords
    from nltk.tokenize import word_tokenize
    ```

2. Load the `.csv` into a dataframe using `read_csv`:

    ```
    Reviews_data = pd.read_csv("data/cleaned_reviews_data.csv")
    ```

3. Check out the first five rows using the `head` method:

    ```
    Reviews_data.head()
    ```

	Review	Liked	reviews_cleaned
0	Wow... Loved this place.	1	wow ... loved this place
1	Crust is not good.	0	crust is not good
2	Not tasty and the texture was just nasty.	0	not tasty and the texture was just nasty
3	Stopped by during the late May bank holiday of...	1	stopped by during the late may bank holiday of...
4	The selection on the menu was great and so wer...	1	the selection on the menu was great and so wer...

Figure 8.8: First 5 rows of the cleaned Reviews data

Check the number of columns and rows:

```
reviews_data.shape
(900, 3)
```

4. Tokenize the reviews data using the `word_tokenize` function in nltk:

```
reviews_data['reviews_tokenized'] = reviews_data['reviews_
cleaned'].apply(word_tokenize)
```

5. Create a `combine_words` custom function that combines all words into a list:

```
def combine_words(word_list):
    all_words = []
    for word in word_list:
        all_words += word
    return all_words
```

6. Create a `count_topwords` custom function that counts the frequency of each word and returns the top 20 words based on the frequency:

```
def count_topwords(all_words):
    counts = dict()
    for word in all_words:
        if word in counts:
            counts[word] += 1
        else:
            counts[word] = 1

    word_count = pd.DataFrame([counts])
    word_count_transposed = word_count.T.reset_index()
    word_count_transposed.columns = ['words','word_count']
    word_count_sorted = word_count_transposed.sort_values("word_
count",ascending = False)
    word_count_sorted
    return word_count_sorted[:20]
```

7. Apply the `combine_words` custom function to the tokenized reviews data to get a list of all words:

```
reviews = reviews_data['reviews_tokenized']
reviews_words =  combine_words(reviews)
reviews_words[:10]
['wow', '...', 'loved', 'this', 'place', 'crust', 'is', 'not',
'good', 'not']
```

8.   Apply the `count_topwords` custom function to create a dataframe with the top 20 words:

```
reviews_topword_count = count_topwords(reviews_words)
reviews_topword_count.head()
```

This results in the following output:

	words	word_count
11	the	517
10	and	360
36	i	310
13	was	267
66	a	208

Figure 8.9: First 5 rows of the top most frequent words

9.   Plot a bar plot of the top words (stop words included) using the `barplot` function in `seaborn`:

```
plt.figure(figsize= (18,10))

sns.barplot(data = reviews_topword_count ,x= reviews_topword_
count['words'], y= reviews_topword_count['word_count'] )
```

This results in the following output:

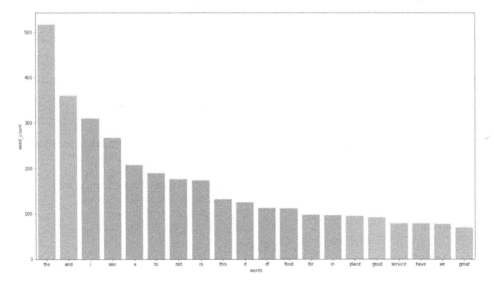

Figure 8.10: Bar plot showing top 20 words based on word count (stop words included)

10. Remove stop words from the reviews data using the `words` attribute within the `stopwords` module in `nltk`:

```
stop_words = set(stopwords.words('english'))
reviews_data['reviews_no_stopwords'] = reviews_data['reviews_
tokenized'].apply(lambda x: [word for word in x if word not in
stop_words])
reviews_data.head()
```

This results in the following output:

	Review	Liked	reviews_cleaned	reviews_tokenized	reviews_no_stopwords
0	Wow... Loved this place.	1	wow ... loved this place	[wow, ..., loved, this, place]	[wow, ..., loved, place]
1	Crust is not good.	0	crust is not good	[crust, is, not, good]	[crust, good]
2	Not tasty and the texture was just nasty	0	not tasty and the texture was just nasty	[not, tasty, and, the, texture, was, just, nasty]	[tasty, texture, nasty]
3	Stopped by during the late May bank holiday of...	1	stopped by during the late may bank holiday of...	[stopped, by, during, the, late, may, bank, ho...	[stopped, late, may, bank, holiday, rick, stev...
4	The selection on the menu was great and so wer...	1	the selection on the menu was great and so wer...	[the, selection, on, the, menu, was, great, an...	[selection, menu, great, prices]

Figure 8.11: First 5 rows of the Reviews data without stopwords

11. Convert the values within the `reviews_no_stopwords` column from a list to a string.

```
reviews_data['reviews_cleaned_stopwords'] = [' '.join(l) for l
in reviews_data['reviews_no_stopwords']]
reviews_data.head()
```

This results in the following output:

	Review	Liked	reviews_cleaned	reviews_tokenized	reviews_no_stopwords	reviews_cleaned_stopwords
0	Wow... Loved this place.	1	wow ... loved this place	[wow, ..., loved, this, place]	[wow, ..., loved, place]	wow ... loved place
1	Crust is not good.	0	crust is not good	[crust, is, not, good]	[crust, good]	crust good
2	Not tasty and the texture was just nasty.	0	not tasty and the texture was just nasty	[not, tasty, and, the, texture, was, just, nasty]	[tasty, texture, nasty]	tasty texture nasty
3	Stopped by during the late May bank holiday of...	1	stopped by during the late may bank holiday of...	[stopped, by, during, the, late, may, bank, ho...	[stopped, late, may, bank, holiday, rick, stev...	stopped late may bank holiday rick steve recom...
4	The selection on the menu was great and so wer...	1	the selection on the menu was great and so wer...	[the, selection, on, the, menu, was, great, an...	[selection, menu, great, prices]	selection menu great prices

Figure 8.12: First 5 rows of the Reviews data without stopwords

12. Apply the `combine_words` custom function to the tokenized reviews data to get a list of all words:

```
reviews_no_stopwords = reviews_data['reviews_no_stopwords']
reviews_words =  combine_words(reviews_no_stopwords)
reviews_words[:10]
['wow',
 '...',
 'loved',
 'place',
 'crust',
 'good',
```

```
'tasty',
'texture',
'nasty',
'stopped']
```

13. Apply the `count_topwords` custom function to create a dataframe with the top 20 words:

```
reviews_topword_count = count_topwords(reviews_words)
reviews_topword_count.head()
```

This displays the following output:

	words	word_count
81	food	112
3	place	95
5	good	92
41	service	79
19	great	70

Figure 8.13: First 5 rows of the most frequent words (stop words excluded)

14. Plot a bar plot of the top words (stop words excluded) using the `barplot` function in `seaborn`:

```
plt.figure(figsize= (18,10))
sns.barplot(data = reviews_topword_count ,x= reviews_topword_
count['words'], y= reviews_topword_count['word_count'] )
```

This results in the following output:

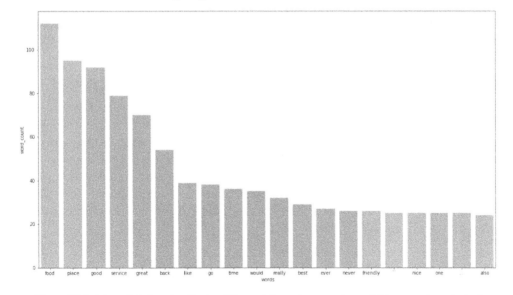

Figure 8.14: A bar plot showing the top 20 words based on word count (stop words excluded)

Awesome! We just removed stop words and visualized the words in the reviews (stop words excluded).

How it works...

In *step 1*, we import the relevant libraries such as `pandas`, `matplotlib`, `seaborn`, and `nltk`. In *step 2*, we load the output `.csv` file from the previous recipe into a `pandas` dataframe using `read_csv`.

In *step 3* and *step 4*, we inspect the dataset using the `head` and `shape` attributes. We tokenize the data using the `word_tokenize` function in the `nltk` library. In *step 5*, we create the custom `combine_words` function that creates a list of all words within the dataframe. In *step 6*, we create another custom function, `count_topwords`, that counts the frequency of each word across the dataframe and identifies the top 20 words based on frequency. It takes in the output of the `combine_words` function as its input.

In *step 7* and *step 8*, we apply the `combine_words` custom function to the tokenized reviews data to get a list of all the words/tokens in the dataframe. We then apply the `count_topwords` custom function to create a dataframe with the top 20 words. In *step 9*, we create a bar plot of the top 20 words using the `barplot` function in `seaborn`. As you may have noticed, many of the top words are stop words, which do not convey significant meaning in the text.

In *step 10 and 11*, we remove the stop words from the reviews data. We get a unique list of stop words from the `nltk stopwords` module using the `words` attribute. We then pass this into a lambda function and apply the function on all the rows within the dataframe, using the `apply` method in `pandas`. We then convert the output from a list to a string using the join method in a lambda function.

In *step 12* to *step 14*, we apply the `combine_words` custom function to the reviews data without stop words. We also apply the `count_topwords` custom function to create a dataframe with the top 20 words, in which stop words are excluded. Lastly, we create a bar plot of the top 20 words (stop words excluded). The output now conveys more meaningful words from our text data, since the stop words have been removed.

There's more...

Sometimes, our text may have special characters, punctuations, or stop words that are not contained in the standard libraries for text preprocessing. In this type of case, we can define these characters or letters in a list variable and apply a custom function to eliminate them. A good example of this instance can be seen in the final chart in *Figure 8.14*. Even though stop words and punctuations have been removed, we still have the ... and . . characters. To eliminate these, we can use the same function used to remove stop words in this recipe. However, we will need to replace our `stopwords` variable with a variable containing a list of the characters to be removed.

Analyzing part of speech

Part of speech analysis involves identifying and tagging the part of speech in a given text. This helps to understand the grammatical structure of the words in our text, which can be very useful to extract relevant insights. For example, using part of speech analysis, we can easily find key terms such as nouns, verbs, or adjectives within our text. These key terms can be pointers to key events, names, products, services, places, descriptive words, and so on.

There are nine parts of speech in the English language; we will focus on the most critical four.

- **Nouns**: These are used to name and identify people, places, or things – for example, a cat, ball, or London.

- **Verbs**: These are used to express actions or occurrences. They go hand in hand with nouns – for example, run, jump, or sleep.

- **Adjectives**: These provide additional information about nouns or pronouns –for example, quick fox, brown shoes, or lovely city.

- **Adverbs**: These provide additional information about verbs – for example, run quickly or speak clearly.

Let's look at some examples:

- The boy ran quickly

- Lagos is a beautiful city

Now, let's do some part of speech tagging:

- *Boy* – noun, *ran* – verb, and *quickly* – adverb

- *Lagos* – noun, *is* – verb, *beautiful* – adjective, and *city* – noun

We will explore part of speech tagging using the `pos_tag` function in `nltk`.

Getting ready

We will work with the Sentiment Analysis of Restaurant Review dataset for this recipe. However, we will work with the preprocessed version that was exported from the *Removing Stop words* recipe. This version cleaned out stop words, punctuations, contractions, uppercase letters, and so on. You can retrieve all the files from the GitHub repository.

How to do it...

We will learn how to prepare and identify parts of speech within our text data using libraries such as `nltk`, `seaborn`, `matplotlib`, and `pandas`:

1.   Import the relevant libraries:

```
import pandas as pd
import matplotlib.pyplot as plt
import seaborn as sns
import nltk
from nltk.tokenize import word_tokenize
from nltk import pos_tag
```

2.   Load the `.csv` into a dataframe using `read_csv`:

```
reviews_data = pd.read_csv("data/cleaned_reviews_no_stopwords_
data.csv")
```

3.   Check the first five rows using the `head` method:

```
reviews_data.head()
```

This displays the following output:

	Review	Liked	reviews_cleaned	reviews_cleaned_stopwords
0	Wow... Loved this place.	1	wow ... loved this place	wow ... loved place
1	Crust is not good.	0	crust is not good	crust good
2	Not tasty and the texture was just nasty.	0	not tasty and the texture was just nasty	tasty texture nasty
3	Stopped by during the late May bank holiday of...	1	stopped by during the late may bank holiday of...	stopped late may bank holiday rick steve recom...
4	The selection on the menu was great and so wer...	1	the selection on the menu was great and so wer...	selection menu great prices

Figure 8.15: First 5 rows of the cleaned Reviews data

Check the number of columns and rows:

```
reviews_data.shape
(900, 4)
```

4.   Remove rows containing missing values using the `dropna` method in `pandas`:

```
reviews_data = reviews_data.dropna()
```

5.   Tokenize the reviews data using the `word_tokenize` function in `nltk`:

```
reviews_data['reviews_tokenized'] = reviews_data['reviews_
cleaned_stopwords'].apply(word_tokenize)
```

6.   Create part of speech tags for the review data using the `pos_tags` function in the `tag` module of `nltk`:

```
reviews_data['reviews_pos_tags'] = reviews_data['reviews_
tokenized'].apply(nltk.tag.pos_tag)
reviews_data.head()
```

This results in the following output:

	Review	Liked	reviews_cleaned	reviews_cleaned_stopwords	reviews_tokenized	reviews_pos_tags
0	Wow... Loved this place.	1	wow ... loved this place	wow ... loved place	[wow, ..., loved, place]	[(wow, NN), (..., :), (loved, VBD), (place, NN)]
1	Crust is not good	0	crust is not good	crust good	[crust, good]	[(crust, NN), (good, NN)]
2	Not tasty and the texture was just nasty.	0	not tasty and the texture was just nasty	tasty texture nasty	[tasty, texture, nasty]	[(tasty, JJ), (texture, NN), (nasty, NN)]
3	Stopped by during the late May bank holiday of...	1	stopped by during the late may bank holiday of...	stopped late may bank holiday rick steve recom...	[stopped, late, may, bank, holiday, rick, stev...	[(stopped, VBN), (late, JJ), (may, MD), (bank,...
4	The selection on the menu was great and so wer...	1	the selection on the menu was great and so wer...	selection menu great prices	[selection, menu, great, prices]	[(selection, NN), (menu, VBZ), (great, JJ), (p...

Figure 8.16: First 5 rows of the Reviews data with pos tags

7. Extract adjectives within the reviews data using a list comprehension:

```
reviews_data['reviews_adjectives'] = reviews_data['reviews_pos_
tags'].apply(lambda x: [word for (word, pos_tag) in x if 'JJ' in
(word, pos_tag)])
```

8. Create a `combine_words` custom function that combines all words into a list:

```
def combine_words(word_list):
    all_words = []
    for word in word_list:
        all_words += word
    return all_words
```

9. Create a `count_topwords` custom function that counts the frequency of each word and returns the top 20 words based on frequency:

```
def count_topwords(all_words):
    counts = dict()
    for word in all_words:
        if word in counts:
            counts[word] += 1
        else:
            counts[word] = 1

    word_count = pd.DataFrame([counts])
    word_count_transposed = word_count.T.reset_index()
    word_count_transposed.columns = ['words','word_count']
    word_count_sorted = word_count_transposed.sort_values("word_
count",ascending = False)
    word_count_sorted
    return word_count_sorted[:20]
```

10. Apply the `combine_words` custom function to the adjectives in the reviews data to get a list of all adjectives:

```
reviews = reviews_data['reviews_adjectives']
reviews_words =  combine_words(reviews)
reviews_words[:10]
['tasty',
 'late',
 'rick',
 'great',
 'angry',
 'fresh',
 'great',
 'great',
 'tried',
 'pretty']
```

11. Apply the `count_topwords` function to the list of adjectives to get the top adjectives:

```
reviews_topword_count = count_topwords(reviews_words)
reviews_topword_count.head()
```

This results in the following output:

	words	word_count
17	good	87
3	great	70
54	delicious	23
44	nice	21
38	bad	15

Figure 8.17: First 5 rows of the most frequent adjectives

12. Plot the top adjectives in a bar plot using the `barplot` function in `seaborn`:

```
plt.figure(figsize= (18,10))

sns.barplot(data = reviews_topword_count ,x= reviews_topword_
count['words'], y= reviews_topword_count['word_count'] )
```

This results in the following output:

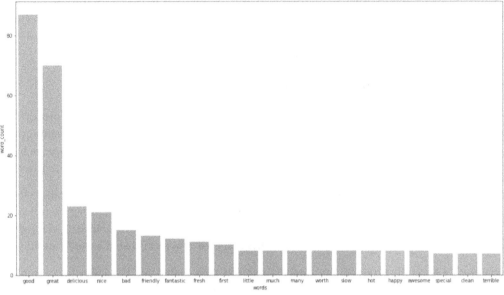

Figure 8.18: A bar plot showing the top 20 adjectives based on word count

Great. We just tagged our reviews with the relevant part of speech and visualized all the adjectives.

How it works...

In *step 1* and *step 2*, we import the relevant libraries and load our reviews data from a `.csv` file. The `.csv` file is an output of the *Removing Stop words* recipe; it contains the reviews data with stop words excluded.

From *step 3* to *step 5*, we inspect our data using the `head` and `shape` attributes. We also drop missing values using the `dropna` method in `pandas`. This is a requirement for tokenization and other preprocessing steps. Even though the original reviews data has no missing values, the processed data with stop words removed has some missing values because some reviews were composed only of stop words, and these became missing values when we removed stop words in the *Removing Stop words* recipe. Lastly, we tokenize the reviews data to break sentences into words.

In *step 6*, we perform part of speech tagging using the `pos_tags` function in the `tag` module of `nltk`. The output of this produces each word and its corresponding part of speech tag in a tuple. Some common tags are tags that start with *J* for adjectives, tags that start with *V* for verbs, tags that start with *N* for nouns, and tags that start with *R* for adverbs. In *step 7*, we extract only adjectives using an `if` statement in a lambda function and apply the function to all the rows in the dataframe.

From *step 8* to *step 12*, we create the custom `combine_words` function to combine all words into a list. We also create another custom `count_topwords` function to get the frequency of each word and identify the top 20 words based on frequency. We then apply the `combine_words` and `count_topwords` custom functions to get a list of all adjectives and the top 20 adjectives based on frequency. Lastly, we create a bar plot of the top 20 adjectives using the `barplot` function in `seaborn`.

Performing stemming and lemmatization

When analyzing text data, we usually need to reduce words to their root or base form. This process is called stemming. Stemming is required because words can appear in several variations depending on the context. Stemming ensures the words are reduced to a common form. This helps to improve the accuracy of our analysis because several variations of the same word can cause noise within our dataset.

Lemmatization also reduces a word to its base or root form; however, unlike stemming, it considers the context and part of speech to achieve this. While stemming just takes off the last characters or suffixes of a word in order to get the root form, lemmatization considers the structure and parts of words, such as root, prefixes, and suffixes, as well as how parts of speech or context change a word's meaning.

Lemmatization generally produces more accurate results than stemming. The following example illustrates this:

- **Original text**: "The food is tasty"
- **Stemmed text**: "The food is tasti"
- **Lemmatized text**: "The food is tasty"

From the preceding example, we can see that stemming performs well on all the words but "tasty." However, we see that lemmatization performs well across all words.

Lemmatization works better when we tag the parts of speech for each word in our text. In some cases, it may produce reasonable results without specifying the part of speech. However, it's a good practice to specify the part of speech for better results.

We will explore the stemming and lemmatization using the `PorterStemmer` and `WordNetLemmatizer` classes in `nltk`.

Getting ready

We will work with the *Sentiment Analysis of Restaurant Review* dataset for this recipe. However, we will work with the preprocessed version, which was exported from the *Removing Stop words* recipe. This version cleaned out stop words, punctuations, contractions, and uppercase letters. You can retrieve all the files from the GitHub repository.

How to do it...

We will learn how to stem and lemmatize our text data using libraries such as `nltk`, `seaborn`, `matplotlib`, and `pandas`:

1. Import the relevant libraries:

```
import pandas as pd
import matplotlib.pyplot as plt
import seaborn as sns
import nltk
from nltk.tokenize import word_tokenize
from nltk.corpus import wordnet
from nltk import pos_tag
from nltk.stem import PorterStemmer
from nltk.stem import WordNetLemmatizer
from nltk.corpus import stopwords
```

2. Load the `.csv` into a dataframe using `read_csv`:

```
reviews_data = pd.read_csv("data/cleaned_reviews_no_stopwords_
data.csv")
```

3. Check the first five rows using the `head` method:

```
reviews_data.head()
```

This displays the following output:

	Review	Liked	reviews_cleaned	reviews_cleaned_stopwords
0	Wow... Loved this place.	1	wow ... loved this place	wow ... loved place
1	Crust is not good.	0	crust is not good	crust good
2	Not tasty and the texture was just nasty.	0	not tasty and the texture was just nasty	tasty texture nasty
3	Stopped by during the late May bank holiday of...	1	stopped by during the late may bank holiday of...	stopped late may bank holiday rick steve recom...
4	The selection on the menu was great and so wer...	1	the selection on the menu was great and so wer...	selection menu great prices

Figure 8.19: First 5 rows of the cleaned Reviews data

Check the number of columns and rows:

```
reviews_data.shape
(900, 4)
```

4. Remove rows containing missing values using the `dropna` method in `pandas`:

```
reviews_data = reviews_data.dropna()
```

5. Tokenize the reviews data using the `word_tokenize` function in `nltk`:

```
reviews_data['reviews_tokenized'] = reviews_data['reviews_
cleaned_stopwords'].apply(word_tokenize)
```

6. Perform stemming on the reviews data using the `PorterStemmer` class in `nltk`:

```
stemmer = nltk.PorterStemmer()
reviews_data['reviews_stemmed_data'] = reviews_data['reviews_
tokenized'].apply(lambda x: [stemmer.stem(word) for word in x])
reviews_data.head()
```

This results in the following output:

	Review	Liked	reviews_cleaned	reviews_cleaned_stopwords	reviews_tokenized	reviews_stemmed_data
0	Wow... Loved this place.	1	wow ... loved this place	wow ... loved place	[wow, ..., loved, place]	[wow, ..., love, place]
1	Crust is not good.	0	crust is not good	crust good	[crust, good]	[crust, good]
2	Not tasty and the texture was just nasty.	0	not tasty and the texture was just nasty	tasty texture nasty	[tasty, texture, nasty]	[tasti, textur, nasti]
3	Stopped by during the late May bank holiday of...	1	stopped by during the late may bank holiday of...	stopped late may bank holiday rick steve recom...	[stopped, late, may, bank, holiday, rick, stev...	[stop, late, may, bank, holiday, rick, steve, ...
4	The selection on the menu was great and so wer...	1	the selection on the menu was great and so wer...	selection menu great prices	[selection, menu, great, prices]	[select, menu, great, price]

Figure 8.20: First 5 rows of the stemmed Reviews data

7. Perform part of speech tagging on the reviews data in preparation for lemmatization. The `pos_tag` function in the `tag` module in `nltk` is used:

```
reviews_data['reviews_pos_tags'] = reviews_data['reviews_
tokenized'].apply(nltk.tag.pos_tag)
reviews_data.head()
```

This results in the following output:

	Review	Liked	reviews_cleaned	reviews_cleaned_stopwords	reviews_tokenized	reviews_pos_tags
0	Wow... Loved this place.	1	wow ... loved this place	wow ... loved place	[wow, ..., loved, place]	[(wow, NN), (..., :), (loved, VBD), (place, NN)]
1	Crust is not good.	0	crust is not good	crust good	[crust, good]	[(crust, NN), (good, NN)]
2	Not tasty and the texture was just nasty.	0	not tasty and the texture was just nasty	tasty texture nasty	[tasty, texture, nasty]	[(tasty, JJ), (texture, NN), (nasty, NN)]
3	Stopped by during the late May bank holiday of...	1	stopped by during the late may bank holiday of...	stopped late may bank holiday rick steve recom...	[stopped, late, may, bank, holiday, rick, stev...	[(stopped, VBN), (late, JJ), (may, MD), (bank,...
4	The selection on the menu was great and so wer...	1	the selection on the menu was great and so wer...	selection menu great prices	[selection, menu, great, prices]	[(selection, NN), (menu, VBZ), (great, JJ), (p...

Figure 8.21: First 5 rows of the Reviews data with pos tags

8. Create a custom function that converts `pos_tag` tags to the `wordnet` format to aid lemmatization. Apply the function to the reviews data:

```
def get_wordnet_pos(tag):
    if tag.startswith('J'):
```

```
            return wordnet.ADJ
        elif tag.startswith('V'):
            return wordnet.VERB
        elif tag.startswith('N'):
            return wordnet.NOUN
        elif tag.startswith('R'):
            return wordnet.ADV
        else:
            return wordnet.NOUN

    reviews_data['reviews_wordnet_pos_tags'] = reviews_
    data['reviews_pos_tags'].apply(lambda x: [(word, get_wordnet_
    pos(pos_tag)) for (word, pos_tag) in x])
    reviews_data.head()
```

This results in the following output

	Review	Liked	reviews_cleaned	reviews_cleaned_stopwords	reviews_tokenized	reviews_stemmed_data	reviews_pos_tags	reviews_wordnet_pos_tags
0	Wow... Loved this place.	1	wow ... loved this place	wow ... loved place	[wow, ..., loved, place]	[wow, ..., love, place]	[(wow, NN), (..., :), (loved, VBD), (place, NN)]	[(wow, n), (..., n), (loved, v), (place, n)]
1	Crust is not good.	0	crust is not good	crust good	[crust, good]	[crust, good]	[(crust, NN), (good, NN)]	[(crust, n), (good, n)]
2	Not tasty and the texture was just nasty.	0	not tasty and the texture was just nasty	tasty texture nasty	[tasty, texture, nasty]	[tasti, textur, nasti]	[(tasty, JJ), (texture, NN), (nasty, NN)]	[(tasty, a), (texture, n), (nasty, n)]
3	Stopped by during the late May bank holiday of...	1	stopped by during the late may bank holiday of...	stopped late may bank holiday rick steve recom...	[stopped, late, may, bank, holiday, rick, stev...	[stop, late, may, bank, holiday, rick, steve, ...	[(stopped, VBN), (late, JJ), (may, MD), (bank,...	[(stopped, v), (late, a), (may, n), (bank, n),...
4	The selection on the menu was great and so wer...	1	the selection on the menu was great and so wer...	selection menu great prices	[selection, menu, great, prices]	[select, menu, great, price]	[(selection, NN), (menu, VBZ), (great, JJ), (p...	[(selection, n), (menu, v), (great, a), (price...

Figure 8.22: First 5 rows of the Reviews data with wordnet pos tags

9. Perform lemmatization using the `lemmatize` method within the `WordNetLemmatizer` class in `nltk`:

```
lemmatizer = WordNetLemmatizer()
reviews_data['reviews_lemmatized'] = reviews_data['reviews_
wordnet_pos_tags'].apply(lambda x: [lemmatizer.lemmatize(word,
tag) for word, tag in x])
reviews_data.head()
```

This displays the following output:

Liked	reviews_cleaned	reviews_cleaned_stopwords	reviews_tokenized	reviews_stemmed_data	reviews_pos_tags	reviews_wordnet_pos_tags	reviews_lemmatized
1	wow ... loved this place	wow ... loved place	[wow, ..., loved, place]	[wow, ..., love, place]	[(wow, NN), (..., :), (loved, VBD), (place, NN)]	[(wow, n), (..., n), (loved, v), (place, n)]	[wow, ..., love, place]
0	crust is not good	crust good	[crust, good]	[crust, good]	[(crust, NN), (good, NN)]	[(crust, n), (good, n)]	[crust, good]
0	not tasty and the texture was just nasty	tasty texture nasty	[tasty, texture, nasty]	[tasti, textur, nasti]	[(tasty, JJ), (texture, NN), (nasty, NN)]	[(tasty, a), (texture, n), (nasty, n)]	[tasty, texture, nasty]
1	stopped by during the late may bank holiday of...	stopped late may bank holiday rick steve recom...	[stopped, late, may, bank, holiday, rick, stev...]	[stop, late, may, bank, holiday, rick, steve, ...]	[(stopped, VBN), (late, JJ), (may, MD), (bank,...]	[(stopped, v), (late, a), (may, n), (bank, n), ...]	[stop, late, may, bank, holiday, rick, steve, ...]
1	the selection on the menu was great and so wer...	selection menu great prices	[selection, menu, great, prices]	[select, menu, great, price]	[(selection, NN), (menu, VBZ), (great, JJ), (p...]	[(selection, n), (menu, v), (great, a), (price...]	[selection, menu, great, price]

Figure 8.23: First 5 rows of the lemmatized Reviews data

10. Convert the values within the `reviews_lemmatized` column from a list to a string:

```
reviews_data['reviews_cleaned_lemmatized'] = [' '.join(l) for l
in reviews_data['reviews_lemmatized']]
reviews_data.head()
```

This results in the following output:

ws_cleaned_stopwords	reviews_tokenized	reviews_stemmed_data	reviews_pos_tags	reviews_wordnet_pos_tags	reviews_lemmatized	reviews_cleaned_lemmatized
wow ... loved place	[wow, ..., loved, place]	[wow, ..., love, place]	[(wow, NN), (..., :), (loved, VBD), (place, NN)]	[(wow, n), (..., n), (loved, v), (place, n)]	[wow, ..., love, place]	wow ... love place
crust good	[crust, good]	[crust, good]	[(crust, NN), (good, NN)]	[(crust, n), (good, n)]	[crust, good]	crust good
tasty texture nasty	[tasty, texture, nasty]	[tasti, textur, nasti]	[(tasty, JJ), (texture, NN), (nasty, NN)]	[(tasty, a), (texture, n), (nasty, n)]	[tasty, texture, nasty]	tasty texture nasty
ed late may bank holiday rick steve recom...	[stopped, late, may, bank, holiday, rick, stev...]	[stop, late, may, bank, holiday, rick, steve, ...]	[(stopped, VBN), (late, JJ), (may, MD), (bank,...]	[(stopped, v), (late, a), (may, n), (bank, n), ...]	[stop, late, may, bank, holiday, rick, steve, ...]	stop late may bank holiday rick steve recommen...
lection menu great prices	[selection, menu, great, prices]	[select, menu, great, price]	[(selection, NN), (menu, VBZ), (great, JJ), (p...]	[(selection, n), (menu, v), (great, a), (price...]	[selection, menu, great, price]	selection menu great price

Figure 8.24: First 5 rows of the lemmatized Reviews data

11. Create a `combine_words` custom function that combines all words into a list:

```
def combine_words(word_list):
    all_words = []
    for word in word_list:
        all_words += word
    return all_words
```

12. Create a `count_topwords` custom function that counts the frequency of each word and returns the top 20 words based on frequency:

```
def count_topwords(all_words):
    counts = dict()
    for word in all_words:
        if word in counts:
            counts[word] += 1
        else:
            counts[word] = 1

    word_count = pd.DataFrame([counts])
    word_count_transposed = word_count.T.reset_index()
    word_count_transposed.columns = ['words','word_count']
    word_count_sorted = word_count_transposed.sort_values("word_count",ascending = False)
    word_count_sorted
    return word_count_sorted[:20]
```

13. Apply the `combine_words` custom function to the lemmatized reviews data to get a list of all lemmatized words:

```
reviews = reviews_data['reviews_lemmatized']
reviews_words =  combine_words(reviews)
reviews_words[:10]
['wow',
 '...',
 'love',
 'place',
 'crust',
 'good',
 'tasty',
 'texture',
 'nasty',
 'stop']
```

14. Apply the count_topwords to the list of all lemmatized words:

```
reviews_topword_count = count_topwords(reviews_words)
reviews_topword_count.head()
```

This results in the following output:

	words	word_count
81	food	113
3	place	99
5	good	97
41	service	79
19	great	71

Figure 8.25: First 5 rows of the most frequent lemmatized words

15. Plot the top lemmatized words using the barplot function in seaborn:

```
plt.figure(figsize= (18,10))

sns.barplot(data = reviews_topword_count ,x= reviews_topword_
count['words'], y= reviews_topword_count['word_count'] )
```

This results in the following output:

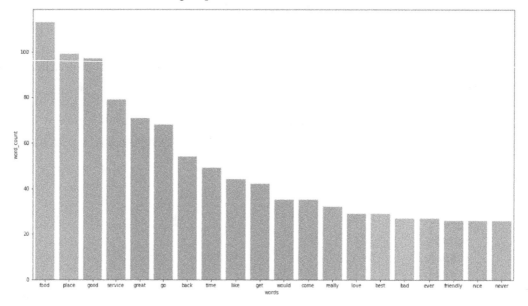

Figure 8.26: A bar plot showing the top 20 lemmatized words based on word count

Awesome! We have now stemmed and lemmatized our data.

How it works...

From *step 1* to *step 2*, we import the relevant libraries and load our reviews data from a `.csv` file. The `.csv` file contains the reviews data with stop words excluded.

From *step 3* to *step 5*, we get a glimpse of the first five rows of our data using `head`, and we check the shape using the `shape` attribute. We then use the `dropna` method to remove missing values caused by the removal of stop words. Lastly, we tokenize the data using the `word_tokenize` function in `nltk`.

In *step 6*, we perform stemming on the reviews data using the `PorterStemmer` class in `nltk`. We use the `stem` method within the `PorterStemmer` class in `nltk` and pass it into a lambda function. We then apply the function on all the rows in the dataframe using the `apply` method in `pandas`.

In *step 7*, we perform part of speech tagging on the tokenized data in preparation for lemmatization. This improves the accuracy of lemmatization because it provides better context. In *step 8*, we then create a custom function that converts the tags to the `wordnet` equivalent, which is what our lemmatizer works with. In *step 9*, we lemmatize the reviews data using the `lemmatize` method within the `WordNetLemmatizer` class. We use a lambda function and the `apply` method in `pandas` to apply the method to all the rows in the dataframe. In *step 10*, we convert the lemmatized values in the dataframe cells from lists to strings.

From *step 11* to *step 15*, we create the `combine_words` custom function to combine all lemmatized words into a list. We also create another custom function, `count_topwords`, to get the frequency of each lemmatized word and identify the top 20 lemmatized words based on frequency. We then apply the `combine_words` and `count_topwords` custom functions to get a list of all adjectives and the top 20 lemmatized words based on frequency. Lastly, we create a bar plot of the top 20 lemmatized words using the `barplot` function in `seaborn`.

Analyzing ngrams

An n-gram is a continuous sequence of *n* items of a given text. These items can be words, letters, or syllables. N-grams help us extract useful information about the distribution of words, syllables, or letters within a given text. The *n* stands for positive numerical values, starting from 1 to *n*. The most common n-grams are unigram, bigram, and trigram, where *n* is 1, 2, and 3 respectively.

Analyzing n-grams involves checking the frequency or distribution of an n-gram within a text. We typically split the text into the respective n-gram and count the frequency of each one in the text data. This will help us identify the most common words, syllables, or phrases in our data.

For example, in the sentence "The boy threw the ball," the n-grams would be as follows:

- 1-gram (or unigram): `["The", "boy", "threw", "the", "ball"]`
- 2-gram (or bigram): `["The boy", "boy threw", "threw the", "the ball"]`

- 3-gram (or trigram): `["The boy threw", "boy threw the", "threw the ball"]`

We will explore n-gram analysis using the `ngram` function in `nltk`.

Getting ready

We will work with the *Sentiment Analysis of Restaurant Review* dataset for this recipe. However, we will work with the preprocessed version that was exported from the *Removing stop words* recipe. This version cleaned out stop words, punctuations, contractions, and uppercase letters. You can retrieve all the files from the GitHub repository.

How to do it...

We will learn how to create and visualize n-grams for our text data using libraries such as `nltk`, `seaborn`, `matplotlib`, and `pandas`:

1. Import the relevant libraries:

    ```
    import pandas as pd
    import matplotlib.pyplot as plt
    import seaborn as sns
    import nltk
    from nltk.tokenize import word_tokenize
    from nltk import ngrams
    ```

2. Load the `.csv` into a dataframe using `read_csv`:

    ```
    reviews_data = pd.read_csv("data/cleaned_reviews_no_stopwords_
    data.csv")
    ```

3. Check the first five rows using the `head` method:

    ```
    reviews_data.head()
    ```

    This displays the following output:

	Review	Liked	reviews_cleaned	reviews_cleaned_stopwords
0	Wow... Loved this place.	1	wow ... loved this place	wow ... loved place
1	Crust is not good.	0	crust is not good	crust good
2	Not tasty and the texture was just nasty.	0	not tasty and the texture was just nasty	tasty texture nasty
3	Stopped by during the late May bank holiday of...	1	stopped by during the late may bank holiday of...	stopped late may bank holiday rick steve recom...
4	The selection on the menu was great and so wer...	1	the selection on the menu was great and so wer...	selection menu great prices

Figure 8.27: First 5 rows of the cleaned Reviews data

Check the number of columns and rows:

```
reviews_data.shape
(900, 4)
```

4. Remove rows containing missing values using the dropna method in pandas :

```
reviews_data = reviews_data.dropna()
```

5. Tokenize the reviews data using the word_tokenize function in nltk:

```
reviews_data['reviews_tokenized'] = reviews_data['reviews_
cleaned_stopwords'].apply(word_tokenize)
```

6. Create a custom function that extracts n-grams from text data using the ngrams function in nltk:

```
def extract_ngrams(tokenized_data,n):
    ngrams_list = list(nltk.ngrams(tokenized_data, n))
    ngrams_str = [' '.join(grams) for grams in ngrams_list]
    return ngrams_str
```

7. Apply the custom function to extract the bigrams in the text:

```
reviews_data['reviews_ngrams'] = reviews_data['reviews_
tokenized'].apply(lambda x: extract_ngrams(x, 2))
reviews_data.head()
```

This results in the following output:

	Review	Liked	reviews_cleaned	reviews_cleaned_stopwords	reviews_tokenized	reviews_ngrams
0	Wow... Loved this place.	1	wow ... loved this place	wow ... loved place	[wow, ..., loved, place]	[wow ... ..., loved, loved place]
1	Crust is not good.	0	crust is not good	crust good	[crust, good]	[crust good]
2	Not tasty and the texture was just nasty.	0	not tasty and the texture was just nasty	tasty texture nasty	[tasty, texture, nasty]	[tasty texture, texture nasty]
3	Stopped by during the late May bank holiday of...	1	stopped by during the late may bank holiday of...	stopped late may bank holiday rick steve recom...	[stopped, late, may, bank, holiday, rick, stev...	[stopped late, late may, may bank, bank holida...
4	The selection on the menu was great and so wer...	1	the selection on the menu was great and so wer...	selection menu great prices	[selection, menu, great, prices]	[selection menu, menu great, great prices]

Figure 8.28: Bigrams in the text

8. Create a combine_words custom function that combines all words into a list.

```
def combine_words(word_list):
    all_words = []
    for word in word_list:
        all_words += word
    return all_words
```

9. Create a `count_topwords` custom function that counts the frequency of each word and returns the top 10 words based on frequency (in this case, ngrams):

```python
def count_topwords(all_words):
    counts = dict()
    for word in all_words:
        if word in counts:
            counts[word] += 1
        else:
            counts[word] = 1

    word_count = pd.DataFrame([counts])
    word_count_transposed = word_count.T.reset_index()
    word_count_transposed.columns = ['words','word_count']
    word_count_sorted = word_count_transposed.sort_values("word_
count",ascending = False)
    word_count_sorted
    return word_count_sorted[:10]
```

10. Apply the `combine_words` custom function to the ngrams in the reviews data to get a list of all ngrams:

```python
reviews = reviews_data['reviews_ngrams']
reviews_words =  combine_words(reviews)
reviews_words[:10]
['wow ...',
 '... loved',
 'loved place',
 'crust good',
 'tasty texture',
 'texture nasty',
 'stopped late',
 'late may',
 'may bank',
 'bank holiday']
```

11. Apply the `count_topwords` to the list of all ngrams:

```python
reviews_topword_count = count_topwords(reviews_words)
reviews_topword_count.head()
```

This results in the following output:

	words	word_count
36	go back	15
366	good food	8
244	great food	8
1475	food good	7
326	really good	6

Figure 8.29: First 5 rows of the most frequent bigrams

12. Plot the top bigrams using the `barplot` function in `seaborn`:

```
plt.figure(figsize= (18,10))

sns.barplot(data = reviews_topword_count ,x= reviews_topword_
count['words'], y= reviews_topword_count['word_count'] )
```

This results in the following output:

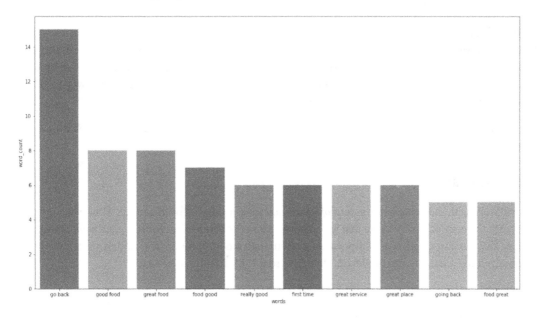

Figure 8.30: A bar plot showing the top 10 bigrams based on count

Good. Now we have extracted bigrams from our review data and visualized them.

How it works...

In *step 1*, we import the pandas, matplotlib, seaborn, and nltk libraries. In *step 2*, we load the reviews data excluding stop words using the read_csv function. From *step 3* to *step 5*, we examine the first five rows using the head method, and we check the shape (the number of rows and columns) using the shape attribute. We use the dropna method to remove missing values. We then tokenize the data to break the sentences into words.

In *step 6*, we create a custom function to extract ngrams from our data. We create the custom function using the ngrams function in nltk . The ngrams function returns a ZIP object containing the ngrams. In the custom function, we convert this ZIP object to a list of tuples and convert this to a list of ngram strings. In *step 7*, we apply the custom function to the reviews data to extract bigrams. We achieve bigrams by providing 2 as the value for the n parameter in the custom function.

From *step 8* to *step 12*, we create the combine_words custom function to combine all ngrams into a list. We also create another custom function, count_topwords, to get the frequency of each ngram and identify the top 10 ngrams based on frequency. We then apply the combine_words and count_topwords custom functions to get a list of all bigrams and the top 10 bigrams based on frequency. Lastly, we create a bar plot of the top 10 bigrams using the barplot function in seaborn.

Creating word clouds

A word cloud is a visual representation of the most common words within a text. It measures the frequency of each word within the text and represents this frequency by the size of the word. Bigger words occur more frequently in the text, while smaller words occur less frequently. A word cloud provides a very useful summary of the word distribution within a text. It is also a great way to gain quick insights into the prominent words in text data.

We will explore how to create word clouds in Python using the wordcloud library and the FreqDist class in nltk.

Getting ready

We will work with the *Sentiment Analysis of Restaurant Review* dataset for this recipe. However, we will work with the preprocessed version that was exported from the *Performing stemming and lemmatization* recipe. This version cleaned out stop words, punctuations, contractions, and uppercase letters and lemmatized the data. You can retrieve all the files from the GitHub repository.

Along with the nltk library, we will also need the wordcloud library:

```
pip install wordcloud
```

How to do it...

We will learn how to create word clouds with our text data using libraries such as `nltk`, `wordcloud`, and `pandas`:

1. Import the relevant libraries:

```
import numpy as np
import matplotlib.pyplot as plt
import pandas as pd
import nltk
from nltk.tokenize import word_tokenize
from nltk.probability import FreqDist
from wordcloud import WordCloud
```

2. Load the `.csv` into a dataframe using `read_csv`:

```
reviews_data = pd.read_csv("data/cleaned_reviews_lemmatized_
data.csv")
```

3. Check the first five rows using the `head` method:

```
reviews_data.head()
```

This displays the following output:

	Review	Liked	reviews_cleaned_lemmatized	reviews_cleaned
0	Wow... Loved this place.	1	wow ... love place	wow ... loved this place
1	Crust is not good.	0	crust good	crust is not good
2	Not tasty and the texture was just nasty.	0	tasty texture nasty	not tasty and the texture was just nasty
3	Stopped by during the late May bank holiday of...	1	stop late may bank holiday rick steve recommen...	stopped by during the late may bank holiday of...
4	The selection on the menu was great and so wer...	1	selection menu great price	the selection on the menu was great and so wer...

Figure 8.31: First 5 rows of the cleaned Reviews data

Check the number of columns and rows:

```
reviews_data.shape
(899, 4)
```

4. Remove rows containing missing values using the `dropna` method in `pandas`:

```
reviews_data = reviews_data.dropna()
```

5. Tokenize the reviews data using the `word_tokenize` function in `nltk`:

```
reviews_data['reviews_tokenized'] = reviews_data['reviews_
cleaned_lemmatized'].apply(word_tokenize)
```

6. Create a `combine_words` custom function that combines all words into a list:

```
def combine_words(word_list):
    all_words = []
    for word in word_list:
        all_words += word
    return all_words
```

7. Apply the `combine_words` custom function to the tokenized reviews data to get a list of all words:

```
reviews = reviews_data['reviews_tokenized']
reviews_words =  combine_words(reviews)
reviews_words[:10]
['wow',
 '...',
 'love',
 'place',
 'crust',
 'good',
 'tasty',
 'texture',
 'nasty',
 'stop']
```

8. Create a word cloud to display the most common words. The `FreqDist` class in `nltk` and the `Wordcloud` class in `wordcloud` are used:

```
mostcommon = FreqDist(reviews_words).most_common(50)
wordcloud = WordCloud(width=1600, height=800, background_
color='white').generate(str(mostcommon))
fig = plt.figure(figsize=(30,10), facecolor='white')
plt.imshow(wordcloud, interpolation="bilinear")
plt.axis('off')
plt.title('Top 100 Most Common Words', fontsize=50)
plt.show()
```

This results in the following output:

Figure 8.32: A word cloud showing the top 50 most common words based on word count

Nice. We have created a word cloud.

How it works...

From *step 1* to *step 2*, we import the relevant libraries and load the `.csv` file containing the lemmatized reviews into a dataframe. From *step 3* to *step 5*, we get a glimpse of the first five rows using the `head` method and check the shape of the data using the `shape` attribute. We then remove missing values using the `dropna` method in `pandas`. We also tokenize the data using the `word_tokenize` function in `nltk`.

From *step 6* to *step 7*, we create the `combine_words` custom function to combine all words into a list. We then apply it to the reviews data to get a list of all words. In *step 8*, we use the `FreqDist` class in `nltk` to identify the most common words. We then use the `WordCloud` class in the `wordcloud` library to create a word cloud. We provide the width, height, and background color arguments to the class. We then use the `figure` function in `matplotlib` to define the figure size and the `imshow` function to display the image generated by the `wordcloud` library. Lastly, we remove the axis using the `axis` function and define a title for the word cloud using the `title` function in `matplotlib`.

Checking term frequency

Term frequency measures the frequency of occurrence of a particular term (word or phrase) in a text. It displays the number of times the term appears in a text relative to the total number of terms in the text. It is very useful to analyze the importance of a term in a text. It is calculated by dividing the number of times the term appears in a text by the total number of terms in that text:

TF = number of times the term appears in a text / total number of terms in the text

The preceding result is a value between 0 and 1 representing the relative frequency of the term in the text.

Term frequency is commonly combined with **Inverse Document Frequency** (**IDF**) to produce a better measure of term relevance or importance. Before discussing IDF, we must first talk about two commonly used concepts we will encounter while analyzing text – documents and corpora (plural of corpus). In simple terms, a corpus is a collection of documents. On the other hand, a document is a unit of text that is usually smaller than a corpus. It can refer to a single text, sentence, article, or paragraph. Let's see an example as follows:

- This is a document

- This is another document

- This is yet another document

- All these documents form a corpus

In the preceding list, each bullet point is a document, while the collection of all bullet points (documents) is a corpus.

IDF measures how rare a term is within a "corpus," while term frequency measures how frequently a term occurs in a "document." When combined, we get an advanced measure of term importance called **Term Frequency-Inverse Document Frequency** (**TF-IDF**). TF-IDF gives a large weight to terms that appear frequently in a document but rarely in the entire corpus, and a low weight to terms that appear frequently in both the document and the corpus. The latter is likely to be articles, conjunctions, and prepositions that do not have significant importance in the document or corpus. IDF is calculated by dividing the number of documents in the corpus by the number of documents in the corpus containing the term, which then finds the logarithm of the output. TF-IDF is calculated by multiplying the term frequency by the inverse document frequency:

IDF = log (number of documents in the corpus / number of documents in the corpus containing the term)

TF-IDF = TF x IDF

We will explore how to perform TF-IDF in Python using the `TfidfVectorizer` class in the `scikit-learn` library.

Getting ready

We will continue to work with the lemmatized version of *the Sentiment Analysis of Restaurant Review* dataset for this recipe. This preprocessed version was exported from the *Performing stemming and lemmatization* recipe. This version cleaned out stop words, punctuations, contractions, and uppercase letters and lemmatized the data. You can retrieve all the files from the GitHub repository.

Along with the nltk library, we will require the scikit-learn and wordcloud libraries:

```
pip install scikit-learn
pip install wordcloud
```

How to do it...

We will learn how to create and visualize term frequency using libraries such as sklearn, wordcloud, and pandas:

1. Import the relevant libraries:

    ```
    import pandas as pd
    import matplotlib.pyplot as plt
    import nltk
    from sklearn.feature_extraction.text import TfidfVectorizer
    from nltk.probability import FreqDist
    from wordcloud import WordCloud
    ```

2. Load the .csv into a dataframe using read_csv:

    ```
    reviews_data = pd.read_csv("data/cleaned_reviews_lemmatized_
    data.csv")
    ```

3. Check the first five rows using the head method:

    ```
    reviews_data.head()
    ```

    This displays the following output:

	Review	Liked	reviews_cleaned_lemmatized	reviews_cleaned
0	Wow... Loved this place.	1	wow ... love place	wow ... loved this place
1	Crust is not good.	0	crust good	crust is not good
2	Not tasty and the texture was just nasty.	0	tasty texture nasty	not tasty and the texture was just nasty
3	Stopped by during the late May bank holiday of...	1	stop late may bank holiday rick steve recommen...	stopped by during the late may bank holiday of...
4	The selection on the menu was great and so wer...	1	selection menu great price	the selection on the menu was great and so wer...

Figure 8.33: First 5 rows of the cleaned Reviews data

Check the number of columns and rows:

```
reviews_data.shape
(899, 4)
```

4. Remove rows containing missing values using the `dropna` method in `pandas`:

```
reviews_data = reviews_data.dropna()
```

5. Create a term matrix using the `TfidfVectorizer` class in `sklearn`:

```
tfIdfVectorizer=TfidfVectorizer(use_idf = True)
tfIdf = tfIdfVectorizer.fit_transform(reviews_data['reviews_
cleaned_lemmatized'])
tfIdf_output = pd.DataFrame(tfIdf.
toarray(),columns=tfIdfVectorizer.get_feature_names())
tfIdf_output.head(10)
```

This results in the following output:

	00	10	100	12	15	17	1979	20	2007	23	...	year	yellow	yellowtail	yelpers	yet	yucky	yukon	yum	yummy	zero
0	0.0	0.0	0.0	0.0	0.0	0.0	0.0	0.0	0.0	0.0	...	0.0	0.0	0.0	0.0	0.0	0.0	0.0	0.0	0.0	0.0
1	0.0	0.0	0.0	0.0	0.0	0.0	0.0	0.0	0.0	0.0	...	0.0	0.0	0.0	0.0	0.0	0.0	0.0	0.0	0.0	0.0
2	0.0	0.0	0.0	0.0	0.0	0.0	0.0	0.0	0.0	0.0	...	0.0	0.0	0.0	0.0	0.0	0.0	0.0	0.0	0.0	0.0
3	0.0	0.0	0.0	0.0	0.0	0.0	0.0	0.0	0.0	0.0	...	0.0	0.0	0.0	0.0	0.0	0.0	0.0	0.0	0.0	0.0
4	0.0	0.0	0.0	0.0	0.0	0.0	0.0	0.0	0.0	0.0	...	0.0	0.0	0.0	0.0	0.0	0.0	0.0	0.0	0.0	0.0
5	0.0	0.0	0.0	0.0	0.0	0.0	0.0	0.0	0.0	0.0	...	0.0	0.0	0.0	0.0	0.0	0.0	0.0	0.0	0.0	0.0
6	0.0	0.0	0.0	0.0	0.0	0.0	0.0	0.0	0.0	0.0	...	0.0	0.0	0.0	0.0	0.0	0.0	0.0	0.0	0.0	0.0
7	0.0	0.0	0.0	0.0	0.0	0.0	0.0	0.0	0.0	0.0	...	0.0	0.0	0.0	0.0	0.0	0.0	0.0	0.0	0.0	0.0
8	0.0	0.0	0.0	0.0	0.0	0.0	0.0	0.0	0.0	0.0	...	0.0	0.0	0.0	0.0	0.0	0.0	0.0	0.0	0.0	0.0
9	0.0	0.0	0.0	0.0	0.0	0.0	0.0	0.0	0.0	0.0	...	0.0	0.0	0.0	0.0	0.0	0.0	0.0	0.0	0.0	0.0

Figure 8.34: First 10 rows of the term matrix

6. Aggregate the TF-IDF scores into a single score per feature and sort:

```
tfIdf_total = tfIdf_output.T.sum(axis=1)
tfIdf_total.sort_values(ascending = False)
food          30.857469
good          29.941739
place         27.339000
service       27.164228
great         23.934017
              ...
deep           0.252433
gloves         0.252433
temp           0.252433
```

```
eel              0.210356
description      0.210356
```

7. Create a word cloud to display the most common words based on their TF-IDF scores:

```
wordcloud = WordCloud(width = 3000, height = 2000,background_
color='white',max_words=70)
wordcloud.generate_from_frequencies(tfIdf_total)
plt.figure(figsize=(40, 30))
plt.imshow(wordcloud, interpolation='bilinear')
plt.axis("off")
plt.show()
```

This results in the following output

Figure 8.35: A word cloud showing the top 50 most common words based on TF-IDF scores

Well done! You've created a term matrix using TF-IDF.

How it works...

From *step 1* and *step 2*, we import the relevant libraries and load the lemmatized reviews data from a .csv file.

From *step 3* to *step 5*, we inspect our data using the head and shape attributes. We also drop missing values using the dropna method in pandas. We then create a term matrix using the

`TfidfVectorizer` class in `sklearn`. We first initialize the `TfidfVectorizer` class, and then we fit it on our lemmatized data using the `fit` method. We save the TF-IDF scores in a dataframe using the `DataFrame` function in `pandas`. In the function, we convert the TF-IDF scores to an array using the `toarray` method and define the columns as the feature names, using the `get_feature_names` attribute of the `TfidfVectorizer` class.

In *step 6*, we transpose the dataframe and sum all the TF-IDF scores in order to get a single score for each feature name. This is a key requirement because we need a single score per feature name to generate the word cloud. In *step 8*, we use the `WordCloud` class in the `wordcloud` library to create a word cloud. We provide the width, height, and background color arguments to the class. We use the `generate_from_frequencies` function in the word cloud library to create the word cloud from the TF-IDF scores. We then use the `figure` function in `matplotlib` to define the figure size and the `imshow` function to display the image generated by the `wordcloud` library. Lastly, we remove the axis using the `axis` function and define a title for the word cloud using the `title` function in `matplotlib`.

There's more...

Another related concept to TF-IDF is **Bag of Words** (**BoW**). The BoW approach converts text into fixed-length vectors by counting the number of times a term appears in a text. When creating a BoW, we usually identify a list of unique terms in the document or corpus. We then count the number of times each term appears. This forms a fixed length of vectors. In the BoW approach, the order of the words isn't really considered; instead, the focus is the number of times a word occurs in a document.

The following example explains this further:

- The boy is tall
- The boy went home
- The boy kicked the ball

The list of unique terms in the preceding corpus is {"the", "boy", "is", "tall", "went", "home", "kicked", "ball"}. This is a list of eight terms. To create a BoW model, we count the number of times each term appears, as shown here:

Document	the	boy	is	tall	went	home	kicked	ball
1	1	1	1	1	0	0	0	0
2	1	1	0	0	1	1	0	0
3	1	1	0	0	0	0	1	1

Table 8.1: Bag of words example

From the preceding table, we can see that the documents have been converted to fixed-length vectors with a fixed length of eight. Typically, lemmatization or stemming should be performed before creating a Bow, just to reduce the dimensionality of the BoW model.

BoW is a simpler technique that only counts the frequency of words in a document, while TF-IDF is a more sophisticated technique that considers both the frequency of words in a document and the entire corpus. TF-IDF is a better representation of the relevance/importance of words in a document.

See also

```
https://www.analyticsvidhya.com/blog/2021/07/bag-of-words-vs-tfidf-
vectorization-a-hands-on-tutorial/
```

Checking sentiments

When analyzing text data, we may be interested in understanding and analyzing the sentiments conveyed in text. Sentiment analysis helps us to achieve this. It helps to identify the emotional tone expressed within our text. This emotional tone or expressed sentiment is typically classified as a positive, negative, or neutral sentiment. Sentiment analysis is very useful because it typically yields actionable insights about a specific topic that our text covers.

Let's check out some examples:

*   I really like the dress
*   I didn't get value for my money
*   I paid for the dress

By analyzing each text, we can easily identify that the first text expresses a positive sentiment, the second expresses a negative sentiment, and the last text expresses a neutral sentiment. In terms of the insights gleaned, the first points to a satisfied customer, while the second points to a dissatisfied customer who likely needs intervention. When performing sentiment analysis on a corpus, we can classify the sentiments into positive, negative, and neutral categories and further visualize the word distribution of each sentiment category.

We will use the `SentimentIntensityAnalyzer` class in `nltk` to explore sentiment analysis. This class uses the **Valence Aware Dictionary and sEntiment Reasoner** (**VADER**) lexicon and rule-based system for sentiment analysis.

Getting ready

We will continue to work with the lemmatized version of the *Sentiment Analysis of Restaurant Review* dataset for this recipe. This preprocessed version was exported from the *Performing stemming and lemmatization* recipe. This version has cleaned out stop words, punctuations, contractions, and uppercase letters and lemmatized the data. You can retrieve all the files from the GitHub repository.

How to do it...

We will learn how to create and visualize term frequency using libraries such as `nltk`, `wordcloud`, `matplotlib`, and `pandas`:

1. Import the relevant libraries:

    ```python
    import pandas as pd
    import matplotlib.pyplot as plt
    import seaborn as sns
    import nltk
    from nltk.tokenize import word_tokenize
    from wordcloud import WordCloud
    from nltk.probability import FreqDist
    from nltk.sentiment.vader import SentimentIntensityAnalyzer
    ```

2. Load the `.csv` into a dataframe using `read_csv`:

    ```python
    reviews_data = pd.read_csv("data/cleaned_reviews_lemmatized_
    data.csv")
    ```

3. Check the first five rows using the `head` method:

    ```python
    reviews_data.head()
    ```

    This displays the following output:

	Review	Liked	reviews_cleaned_lemmatized	reviews_cleaned
0	Wow... Loved this place.	1	wow ... love place	wow ... loved this place
1	Crust is not good.	0	crust good	crust is not good
2	Not tasty and the texture was just nasty.	0	tasty texture nasty	not tasty and the texture was just nasty
3	Stopped by during the late May bank holiday of...	1	stop late may bank holiday rick steve recommen...	stopped by during the late may bank holiday of...
4	The selection on the menu was great and so wer...	1	selection menu great price	the selection on the menu was great and so wer...

Figure 8.36: First 5 rows of the cleaned Reviews data

4. Check the number of columns and rows:

    ```python
    reviews_data.shape
    (899, 4)
    ```

5. Check sentiments in the review data using the `SentimentIntensityAnalyzer` class in `nltk`.

    ```python
    sentimentanalyzer = SentimentIntensityAnalyzer()
    reviews_data['sentiment_scores'] = reviews_data['reviews_
    cleaned'].apply(lambda x: sentimentanalyzer.polarity_scores(x))
    reviews_data.head()
    ```

This results in the following output:

	Review	Liked	reviews_cleaned_lemmatized	reviews_cleaned	sentiment_scores
0	Wow... Loved this place.	1	wow ... love place	wow ... loved this place	{'neg': 0.0, 'neu': 0.28, 'pos': 0.72, 'compou...
1	Crust is not good.	0	crust good	crust is not good	{'neg': 0.445, 'neu': 0.555, 'pos': 0.0, 'comp...
2	Not tasty and the texture was just nasty.	0	tasty texture nasty	not tasty and the texture was just nasty	{'neg': 0.34, 'neu': 0.66, 'pos': 0.0, 'compou...
3	Stopped by during the late May bank holiday of...	1	stop late may bank holiday rick steve recommen...	stopped by during the late may bank holiday of...	{'neg': 0.093, 'neu': 0.585, 'pos': 0.322, 'co...
4	The selection on the menu was great and so wer...	1	selection menu great price	the selection on the menu was great and so wer...	{'neg': 0.0, 'neu': 0.728, 'pos': 0.272, 'comp...

Figure 8.37: First 5 rows of the Reviews data with the combined sentiment scores

6. Store the overall sentiment scores:

```
reviews_data['overall_scores'] = reviews_data['sentiment_
scores'].apply(lambda x: x['compound'])
reviews_data.head()
```

This displays the following output:

	Review	Liked	reviews_cleaned_lemmatized	reviews_cleaned	sentiment_scores	overall_scores
0	Wow... Loved this place.	1	wow ... love place	wow ... loved this place	{'neg': 0.0, 'neu': 0.28, 'pos': 0.72, 'compou...	0.8271
1	Crust is not good.	0	crust good	crust is not good	{'neg': 0.445, 'neu': 0.555, 'pos': 0.0, 'comp...	-0.3412
2	Not tasty and the texture was just nasty.	0	tasty texture nasty	not tasty and the texture was just nasty	{'neg': 0.34, 'neu': 0.66, 'pos': 0.0, 'compou...	-0.5574
3	Stopped by during the late May bank holiday of...	1	stop late may bank holiday rick steve recommen...	stopped by during the late may bank holiday of...	{'neg': 0.093, 'neu': 0.585, 'pos': 0.322, 'co...	0.6908
4	The selection on the menu was great and so wer...	1	selection menu great price	the selection on the menu was great and so wer...	{'neg': 0.0, 'neu': 0.728, 'pos': 0.272, 'comp...	0.6249

Figure 8.38: First 5 rows of the Reviews data with the overall sentiment scores

7. Categorize the sentiment scores into positive (1) and negative (0).

```
reviews_data['sentiment_category'] = 0
reviews_data.loc[reviews_data['overall_scores'] > 0,'sentiment_
category'] = 1
reviews_data.head()
```

This results in the following output:

	Review	Liked	reviews_cleaned_lemmatized	reviews_cleaned	sentiment_scores	overall_scores	sentiment_category
0	Wow... Loved this place.	1	wow ... love place	wow ... loved this place	{'neg': 0.0, 'neu': 0.28, 'pos': 0.72, 'compou...	0.8271	1
1	Crust is not good.	0	crust good	crust is not good	{'neg': 0.445, 'neu': 0.555, 'pos': 0.0, 'comp...	-0.3412	0
2	Not tasty and the texture was just nasty.	0	tasty texture nasty	not tasty and the texture was just nasty	{'neg': 0.34, 'neu': 0.66, 'pos': 0.0, 'compou...	-0.5574	0
3	Stopped by during the late May bank holiday of...	1	stop late may bank holiday rick steve recommen...	stopped by during the late may bank holiday of...	{'neg': 0.093, 'neu': 0.585, 'pos': 0.322, 'co...	0.6908	1
4	The selection on the menu was great and so wer...	1	selection menu great price	the selection on the menu was great and so wer...	{'neg': 0.0, 'neu': 0.728, 'pos': 0.272, 'comp...	0.6249	1

Figure 8.39: First 5 rows of the Reviews data with the sentiment category

8. Compare the predicted sentiments to the actual sentiments using a crosstab, created by the `crosstab` function in `pandas`. This output is commonly known as a confusion matrix and is used to check the accuracy of models:

```
pd.crosstab(reviews_data['Liked'], reviews_data['sentiment_
category'], rownames=['Actual'], colnames=['Predicted'])
```

This results in the following output:

Predicted	0	1
Actual		
0	324	79
1	91	405

Figure 8.40: Confusion matrix of the actual and predicted sentiment scores

9. Create a `combine_words` custom function that combines all words into a list:

```
def combine_words(word_list):
    all_words = []
    for word in word_list:
        all_words += word
    return all_words
```

10. Identify the negative reviews and extract all the words:

```
reviews_data['reviews_tokenized'] = reviews_data['reviews_
cleaned_lemmatized'].apply(word_tokenize)
reviews = reviews_data.loc[reviews_data['sentiment_
category']==0,'reviews_tokenized']
reviews_words =  combine_words(reviews)
reviews_words[:10]
['crust',
 'good',
 'tasty',
 'texture',
 'nasty',
 'get',
 'angry',
 'want',
 'damn',
 'pho']
```

11. Plot a word cloud containing only the negative reviews:

```
mostcommon = FreqDist(reviews_words).most_common(50)
wordcloud = WordCloud(width=1600, height=800, background_
color='white').generate(str(mostcommon))
fig = plt.figure(figsize=(30,10), facecolor='white')
plt.imshow(wordcloud, interpolation="bilinear")
plt.axis('off')
plt.title('Top 100 Most Common Words for negative feedback',
fontsize=50)
plt.show()
```

This results in the following output:

Figure 8.41: A word cloud showing the top 50 most common words in negative sentiments

Well done! We have now analyzed the sentiments of the reviews data.

How it works...

From *step 1* to *step 2*, we import the relevant libraries and load the `.csv` file into a dataframe. In *step 3*, we inspect the dataset using the `head` and `shape` attributes. In *step 4*, we check for the sentiments in the reviews data using the `SentimentIntensityAnalyzer` class in `nltk`. We get the sentiment scores by using the `polarity_scores` method. The output is a dictionary containing a negative polarity score representing the proportion of negative words in the reviews data, a positive polarity score representing the proportion of positive words in the reviews data, a neutral polarity score representing the proportion of neutral words in the reviews data, and a compound score representing the overall score of the sentiment, which ranges from -1 (extremely negative) to +1 (extremely positive).

In *step 5*, we save the compound score in a column. In step 6, we categorize the sentiments into positive and negative, 1 and 0 respectively for simplicity. For better accuracy, we can classify the score into positive, negative and neutral sentiments. In step 7, we create a confusion matrix to get a sense of the accuracy of the sentiments output.

From *step 8* to *step 10*, we create the `combine_words` custom function to combine all words into a list. We extract all the negative sentiments and apply the `combine_words` function to the negative sentiments. We use the `FreqDist` class in `nltk` to identify the most common words in the negative sentiments. We then use the `WordCloud` class in the `wordcloud` library to create a word cloud. We provide the width, height, and background color arguments to the class. We then use the `figure` function in `matplotlib` to define the figure size and the `imshow` function to display the image generated by the word cloud. Lastly, we remove the axis using the `axis` function and define a title for the word cloud using the `title` function in `matplotlib`.

There's more...

Apart from the `nltk` library, there are a few other options for sentiment analysis we could explore:

- **TextBlob**: TextBlob is a popular library for sentiment analysis. It provides a pre-trained model that can be used to analyze the sentiment of a piece of text and classify it into positive, negative, or neutral. It uses a rules-based approach for sentiment analysis. It determines the sentiment of a text by using a lexicon of words with predetermined polarities.

- **spaCy**: The spaCy library contains various tools for sentiment analysis. It can be used when working on large-scale sentiment analysis tasks that require speed and efficiency.

- **Classification models**: We can use a machine learning classification model for sentiment analysis. To achieve this, we will need a labeled dataset. This means that the dataset must be annotated with its corresponding sentiment label (e.g., positive, negative, or neutral). The model takes a BoW from the text as input. A very common classification model used is the Naïve Bayes model.

- **Deep learning models**: Deep learning models such as **Convolutional Neural Networks (CNNs)**, **Recurrent Neural Networks (RNNs)**, and especially **Long Short-Term Memory (LSTM)** networks can also be used for sentiment analysis. In addition, Transformers (a type of deep learning model), which achieve state-of-the-art performance, can be used too. We can leverage a pretrained transformer model, since it is difficult to build one from scratch. An advantage of deep learning models for sentiment analysis is that they can extract relevant features automatically from text data. This is very useful, especially when dealing with complex and unstructured text data. The only setback is that these models require large amounts of training data and are more computationally expensive. Just like the classification model approach, this approach also requires labeled data.

See also

You can check out the following useful resource:

```
https://www.analyticsvidhya.com/blog/2022/07/sentiment-analysis-
using-python/
```

Performing Topic Modeling

Topic modeling helps us identify the underlying themes or topics within a text. When applied to a corpus, we can easily identify some of the most common topics or themes it covers. Imagine we have a large set of documents, such as news articles, customer reviews or blog posts. The topic modeling algorithms analyzes the words and patterns within the documents to identify recurring themes or topics. It does this by assuming each topic is characterized by a set of words that frequently occur together. By checking the frequency of words which occur together, the algorithm can identify the most likely topics or themes present in a corpus.

Let's see an example with some documents:

Document 1:

"The government needs to consider implementing policies which will stimulate economic growth. The country needs policies aimed at creating jobs and improving overall infrastructure."

Document 2:

"I enjoyed watching the football match over the weekend. The two teams were contending for the trophy and displayed great skill and teamwork. The winning goal was scored during the last play of the game."

Document 3:

"Technology has changed the way we live and do business. The advancements in technology have significantly transformed many industries. Companies are leveraging innovative technologies to drive productivity and profitability. This is now a requirement to stay competitive."

Based on these documents, the topic modeling algorithm may identify the following combination of keywords:

Document 1:

- **Keywords**: government, policies, infrastructure, economy, jobs, growth
- **Probable Topic**: Politics/Economy

Document 2:

- **Keywords**: football, match, teams, score, win, goal, game
- **Probable Topic**: Sports

Document 3:

- **Keywords**: technology, advancements, innovative, productivity, profitability, industries
- **Probable Topic**: Technology

From the preceding example, the topic modeling algorithm may identify the combination of keywords above. These keywords can help us appropriately name the topics based on the keywords. .

> **Note**
>
> The LDA model in `gensim` involves random initialization of some model parameters, such as topic assignments for words and topic proportions for documents. This means the model generates different results each time it is run. It also means you may not get the exact results in this recipe especially for topics.   To ensure getting the same topic results each time the code is run (reproducibility), we will need to use the `random_state` parameter in the model.

We will use the `gensim` and `pyLDAvis` libraries to perform topic modeling and visualize the topics.

Getting ready

We will continue to work with the lemmatized version of the *Sentiment Analysis of Restaurant Review* dataset for this recipe. This preprocessed version was exported from the *Performing stemming and lemmatization* recipe. This version cleaned out stop words, punctuations, contractions, and uppercase letters and lemmatized the data. You can retrieve all the files from the GitHub repository.

Along with the `nltk` library, we will also need the `gensim` and `pyLDAvis` libraries:

```
pip install gensim
pip install pyLDAvis
```

How to do it...

We will learn how to perform topic modeling on our text data and visualize the output using the `nltk`, `gensim`, `pyLDAvis`, and `pandas` libraries:

1. Import the relevant libraries:

    ```
    import pandas as pd
    import nltk
    from nltk.tokenize import word_tokenize
    import gensim
    from gensim.utils import simple_preprocess
    import gensim.corpora as corpora
    from pprint import pprint
    ```

```
import pyLDAvis.gensim_models
pyLDAvis.enable_notebook()# Visualise inside a notebook
```

2. Load the `.csv` into a dataframe using `read_csv`:

```
reviews_data = pd.read_csv("data/cleaned_reviews_lemmatized_
data.csv")
```

3. Check the first five rows using the `head` method:

```
reviews_data.head()
```

This displays the following output:

	Review	Liked	reviews_cleaned_lemmatized	reviews_cleaned
0	Wow... Loved this place.	1	wow ... love place	wow ... loved this place
1	Crust is not good.	0	crust good	crust is not good
2	Not tasty and the texture was just nasty.	0	tasty texture nasty	not tasty and the texture was just nasty
3	Stopped by during the late May bank holiday of...	1	stop late may bank holiday rick steve recommen...	stopped by during the late may bank holiday of...
4	The selection on the menu was great and so wer...	1	selection menu great price	the selection on the menu was great and so wer...

Figure 8.42: First 5 rows of the cleaned Reviews data

Check the number of columns and rows:

```
reviews_data.shape
(899, 4)
```

4. Tokenize the reviews data using the `word_tokenize` function in `nltk`:

```
reviews_data['reviews_tokenized'] = reviews_data['reviews_
cleaned_lemmatized'].apply(word_tokenize)
```

5. Prepare the reviews data for topic modeling. Convert the token data into a list of lists. Subset to check the first item in the list, subset again to check the first item in the resulting list, and lastly the first 30 characters of that output:

```
data = reviews_data['reviews_tokenized'].values.tolist()
all_reviews_words = list(data)
all_reviews_words[:1][0][:30]
['wow', '...', 'love', 'place']
```

6. Create a dictionary and a corpus. Apply the same subset as before to see the results of the corpus for the preceding item:

```
dictionary = corpora.Dictionary(all_reviews_words)
corpus = [dictionary.doc2bow(words) for words in all_reviews_
words]
# View
```

```
print(corpus[:1][0][:30])
[(0, 1), (1, 1), (2, 1), (3, 1)]
```

7. Create an LDA model using the LdaMulticore class within the models module of gensim:

```
topics = 2
lda_model = gensim.models.
LdaMulticore(corpus=corpus,id2word=dictionary, num_
topics=topics)
model_result = lda_model[corpus]

pprint(lda_model.print_topics()[:4])
[(0,
  '0.016*"service" + 0.013*"food" + 0.011*"go" + 0.010*"great" +
0.009*"back" '
  '+ 0.009*"like" + 0.009*"good" + 0.009*"get" + 0.007*"place" +
0.007*"time"'),
 (1,
  '0.021*"place" + 0.020*"food" + 0.019*"good" + 0.012*"great" +
0.010*"go" + '
  '0.009*"service" + 0.008*"time" + 0.007*"back" +
0.007*"really" + '
  '0.007*"best"')]
```

8. Review the output of the model:

```
print(reviews_data['reviews_tokenized'][0])
model_result[0]
['wow', '...', 'love', 'place']
[(0, 0.14129943), (1, 0.8587006)]
```

9. Plot topics and top words using the pyLDAvis library:

```
lda_visuals = pyLDAvis.gensim_models.prepare(lda_model, corpus,
dictionary)
pyLDAvis.display(lda_visuals)
```

This results in the following output:

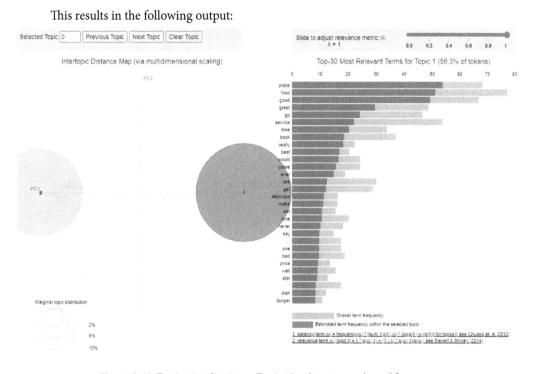

Figure 8.43: Topic visualization – Topic 1's relevant words and frequency

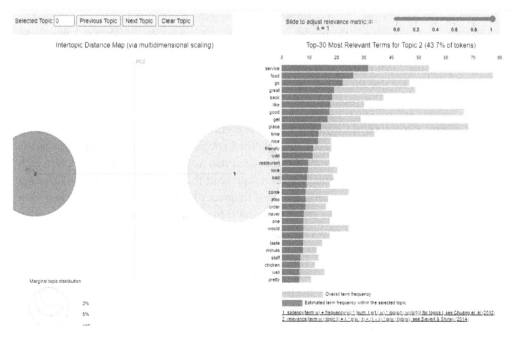

Figure 8.44: Topic visualization – Topic 2's relevant words and frequency

10. Add the output to the original data frame:

```
reviews_data['reviews_topics'] = [sorted(lda_model[corpus]
[i], key=lambda x: x[1], reverse=True)[0][0] for i in
range(len(reviews_data['reviews_tokenized']))]
```

11. Count the number of records in each category using the value_counts method in pandas:

```
reviews_data['reviews_topics'].value_counts()
1    512
0    387
```

Great! We have just performed topic modeling to identify the key themes in our reviews data.

How it works...

From *step 1* to *step 2*, we import the relevant libraries and load the .csv file into a dataframe.

From *step 3* to *step 5*, we inspect the dataset using the head method and shape attribute. We tokenize the reviews data using the word_tokenize function. We then prepare the reviews data for topic modeling by converting the tokens into a list of lists.

In *step 6*, we create a dictionary and a corpus. The corpus is a collection of documents (in this case, each review on each row), while the dictionary is a mapping in which each word in the corpus is assigned a unique ID. In *step 7*, we build an LDA model using the LdaMulticore class within the models module of gensim. We provide the corpus, dictionary, and number of topics as the arguments for the class. We also specify a random_state argument to ensure the result of the code is reproducible. This ensures that each time we run the model, we get the exact same results. We then run the model on our corpus to get the topics. We use the pprint function from the pprint library to print the topics in a pretty format – that is, a more friendly format. In *step 8*, we review the results of the model. From the result, we can see that the review on the first row has an 85.8% probability of being in topic 1 and a 14.1% chance of being in topic 0.

In *step 9*, we display the output of the model using the pyLDAvis library. We use the prepare function from the gensim_models class and pass the LDA model, corpus, and dictionary as the arguments. The chart helps us to visualize the topics and relevant words in each topic. In the chart, each circle represents a topic, and the horizontal bar chart represents the words in the topics. By hovering over a topic, we can view the relevant words; the word frequency is in blue, while the estimated term frequency within the selected topic is in red. A key goal is to ensure that the topics do not overlap because this point to the uniqueness of the topics. Something noteworthy in the chart is that topic 1 focuses on good food, while topic 2 focuses more on good service.

In *step 10*, we append the topics back to the reviews. We use a lambda function to achieve this. The function sorts the list of topics to identify the topic with the highest probability and assigns it to the relevant review. In *step 11*, we use the value_counts method in pandas to count the frequency of the topics in the reviews data.

Choosing an optimal number of topics

To derive the best value from topic modeling, we must choose the optimal number of topics. This can be achieved using a measure of coherence within the topics. Coherence evaluates the quality of the topics by measuring how semantically similar the top words of a topic are. There are various types of coherence measures; however, most of them are based on the calculation of pairwise word co-occurrence statistics. Higher coherence scores typically mean that the topics are more coherent and semantically meaningful.

In `gensim`, we will work with two coherence measures – the `cumulative` coherence (Cumass) and `C_v` coherence. Cumass calculates the pairwise word co-occurrence statistics between the top words in a topic and returns the sum of these scores. Conversly, C_v compares the top words in a topic to a background corpus of words to estimate coherence. It compares the probability of co-occurrence of the top words in the topic to the probability of co-occurrence of the top words in the topic with a set of reference words from the background corpus.

Cumass is a simpler and more efficient measure that is easy to interpret; however, it doesn't handle rare or ambiguous words well as C_v does. Also, it is less computationally expensive than C_v.

We will use the `gensim` library to find the optimal number of topics.

Getting ready

We will continue to work with the lemmatized version of the *Sentiment Analysis of Restaurant Review* dataset for this recipe. This preprocessed version was exported from the *Performing stemming and lemmatization* recipe. This version cleaned out stop words, punctuations, contractions, and uppercase letters and lemmatized the data. You can retrieve all the files from the GitHub repository.

How to do it...

We will learn how to perform topic modeling on our text data and visualize the output using the `nltk`, `gensim`, `pyLDAvis`, and `pandas` libraries:

1. Import the relevant libraries:

    ```
    import pandas as pd
    import nltk
    from nltk.tokenize import word_tokenize
    import gensim
    from gensim.utils import simple_preprocess
    import gensim.corpora as corpora
    from gensim.models import CoherenceModel
    ```

2. Load the `.csv` into a dataframe using `read_csv`:

```
reviews_data = pd.read_csv("data/cleaned_reviews_lemmatized_
data.csv")
```

3. Check the first five rows using the `head` method:

```
reviews_data.head()
```

This displays the following output:

	Review	Liked	reviews_cleaned_lemmatized	reviews_cleaned
0	Wow... Loved this place.	1	wow ... love place	wow ... loved this place
1	Crust is not good.	0	crust good	crust is not good
2	Not tasty and the texture was just nasty.	0	tasty texture nasty	not tasty and the texture was just nasty
3	Stopped by during the late May bank holiday of...	1	stop late may bank holiday rick steve recommen...	stopped by during the late may bank holiday of...
4	The selection on the menu was great and so wer...	1	selection menu great price	the selection on the menu was great and so wer...

Figure 8.45: First 5 rows of the cleaned Reviews data

Check the number of columns and rows:

```
reviews_data.shape
(899, 4)
```

4. Tokenize the reviews data using the `word_tokenize` function in `nltk`:

```
reviews_data['reviews_tokenized'] = reviews_data['reviews_
cleaned_lemmatized'].apply(word_tokenize)
```

5. Prepare the reviews data for topic modeling. Convert the token data into a list of lists. Subset to check the first item in the list, subset again to check the first item in the resulting list, and lastly the first 30 characters of that output.

```
data = reviews_data['reviews_tokenized'].values.tolist()
all_reviews_words = list(data)
all_reviews_words[:1][0][:30]
['wow', '...', 'love', 'place']
```

6. Create a dictionary and a corpus. Apply the same subset as before to see the results of the corpus for the preceding item:

```
dictionary = corpora.Dictionary(all_reviews_words)
corpus = [dictionary.doc2bow(words) for words in all_reviews_
words]
print(corpus[:1][0][:30])
[(0, 1), (1, 1), (2, 1), (3, 1)]
```

7. Perform Umass coherence using the `CoherenceModel` class:

```
review_topics_um = []
coherence_scores_um = []
for i in range(2,10,1):
    lda_model_um = gensim.models.LdaMulticore(corpus=corpus,
id2word=dictionary,num_topics = i ,random_state=1)
    coherence_um = CoherenceModel(model=lda_model_um,
corpus=corpus, dictionary=dictionary, coherence='u_mass')
    review_topics_um.append(i)
    coherence_scores_um.append(coherence_um.get_coherence())
plt.plot(review_topics_um, coherence_scores_um)
plt.xlabel('Number of Topics')
plt.ylabel('Coherence Score')
plt.show()
```

This results in the following output:

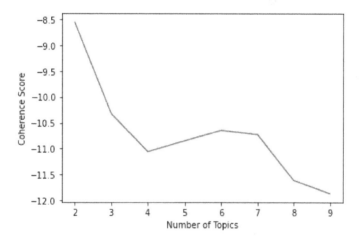

Figure 8.46: A line plot of the Umass coherence scores

8. Identify the optimal number of topics from the Umass coherence scores:

```
topics_range = range(2,10,1)
max_index = coherence_scores_um.index(max(coherence_scores_um))
optimal_num_topics = topics_range[max_index]
print("Optimal topics: ",optimal_num_topics)
```

Optimal topics: 2

9. Perform C_v coherence using the `CoherenceModel` class:

```
review_topics_cv = []
coherence_scores_cv = []
for i in range(2,10,1):
    lda_model_cv = gensim.models.LdaMulticore(corpus=corpus,
id2word=dictionary, num_topics=i, random_state=1)
    coherence_cv = CoherenceModel(model=lda_model_cv,
texts = reviews_data['reviews_tokenized'], corpus=corpus,
dictionary=dictionary, coherence='c_v')
    review_topics_cv.append(i)
    coherence_scores_cv.append(coherence_cv.get_coherence())
plt.plot(review_topics_cv, coherence_scores_cv)
plt.xlabel('Number of Topics')
plt.ylabel('Coherence Score')
plt.show()
```

This results in the following output:

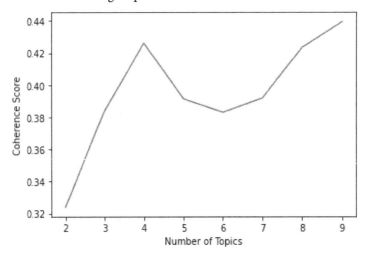

Figure 8.47: A line plot of the C_v coherence scores

10. Identify the optimal number of topics from the C_v coherence scores:

```
topics_range = range(2,10,1)
max_index = coherence_scores_cv.index(max(coherence_scores_cv))
optimal_num_topics = topics_range[max_index]
print("Optimal topics: ",optimal_num_topics)
```

Optimal topics: 9

Good stuff. We have now identified the optimal number of topics.

How it works...

In *step 1*, we import the relevant libraries. In *step 2*, we load the `.csv` file into a dataframe. In *step 3*, we inspect the dataset using the `head` and `shape` attributes. In *step 4*, we tokenize the reviews data using the `word_tokenize` function.

In *step 5*, we prepare the reviews data for topic modeling by converting the tokens into a list of lists. In *step 6*, we create a dictionary and a corpus. In *step 7*, we create a for loop to build several models with a varying number of topics (2 to 9), we build various LDA models on the number of topics range (2,9) using the `LdaMulticore` class in `gensim`. In the same `for` loop, we then compute their Umass coherence score, save the scores in a list, and display the scores in a line chart. We compute the coherence score using the `CoherenceModel` class. In *step 8*, we extract the optimal score for the Umass coherence by looking for the maximum coherence score from the number of topics range (2 to 9). The optimal number of topics is two topics.

From *step 9* to *step 10*, we repeat the preceding steps for the C_v coherence and display the scores in a line chart. As you may have noticed, the different coherence measures give conflicting results. This is because the measures use different approaches to compute coherence. In this case, we will choose the Umass score of two topics because the visualization in the previous recipe shows no overlaps in the topics. On the other hand, the C_v scores proposed optimal topics of 9. These 9 topics have significant overlaps. You can try visualizing both 2 topics and 9 topics using the `pyLDAvis` library to see this.

9

Dealing with Outliers and Missing Values

Outliers and missing values are common issues we will encounter when analyzing various forms of data. They can lead to inaccurate or biased conclusions when not handled properly in our dataset. Hence, it is important to appropriately address them before analyzing our data.

Outliers are unusually high or low values within a dataset that deviate significantly from the rest of the data points in the dataset. Outliers occur due to a wide variety of reasons; the common reasons are covered in this chapter. On the other hand, missing values refer to the absence of data points within a specific variable or observation in our dataset. There are several reasons why they occur; the common reasons are also covered in this chapter.

When handling outliers and missing values, proper care needs to be taken because using the wrong technique can also lead to inaccurate or biased conclusions. An important step when handling missing values and outliers is to identify the reason why they exist within our dataset. This typically provides a pointer to what method will best address them. Other considerations include the nature of the data and the type of analysis to be performed.

In this chapter, we will discuss common techniques for identifying and handling outliers and missing values. The chapter covers the following key recipes:

- Identifying outliers
- Spotting univariate outliers
- Finding bivariate outliers
- Identifying multivariate outliers
- Removing outliers
- Replacing outliers
- Identifying missing values
- Dropping missing values

- Replacing missing values
- Imputing missing values using machine learning models

Technical requirements

We will leverage the `pandas`, `matplotlib`, `seaborn`, `scikit-learn`, `missingno`, and `scipy` libraries in Python for this chapter. The code and notebooks for this chapter are available on GitHub at `https://github.com/PacktPublishing/Exploratory-Data-Analysis-with-Python-Cookbook`.

Identifying outliers

Outliers are unusually high or low values that occur in a dataset. When compared to other observations in a dataset, outliers typically stand out as different and are considered to be extreme values. Some of the reasons outliers occur in a dataset include genuine extreme values, measurement errors, data entry errors, and data processing errors. Measurement errors are typically caused by faulty systems, such as weighing scales, sensors, and so on. Data entry errors occur when inaccurate inputs are provided by users. Examples include mistyping inputs, providing wrong data formats, or swapping values (transposition errors). Processing errors can occur during data aggregation or transformation to generate a final output.

It is very important to spot and handle outliers because they can lead to wrong conclusions and distort any analysis. The following example shows this:

PersonID	Industry	Annual Income
Person 1	Consulting	170,000
Person 2	Tech	500,000
Person 3	Banking	150,000
Person 4	Healthcare	100,000
Person 5	Education	30,000
Person 6	Tech	450,000
Person 7	Banking	350,000
Person 8	Consulting	150,000
Person 9	Healthcare	80,000
Person 10	Education	25,000

Table 9.1: Outlier analysis on income earnings

From the preceding table, the average income, when we consider all figures, is approximately 200,000. It reduces to about 100,000 when we exclude income values above 300,000, and it increases to 130,000 when we further exclude income values below 50,000.

We see that the presence of outliers can significantly influence the average income. When all values are used for the computation, it doesn't give a true reflection of the average because it has been distorted by the outliers present in the dataset (very low and very high values).

To handle outliers appropriately, we need to understand the specific context and consider the objectives of our analysis before concluding the specific approach to use.

In this recipe, we will use the `describe` function in `pandas` and `boxplot` in `seaborn` to identify possible outliers.

Getting ready

We will work with one dataset in this chapter: Amsterdam House Prices Data from Kaggle.

Create a folder specifically for this chapter. Within this folder, create a new Python script or Jupyter Notebook file and also create a data subfolder. Place the `HousingPricesData.csv` file in that subfolder.

Alternatively, you could retrieve all the files from the GitHub repository.

> **Note**
>
> Kaggle provides the Amsterdam House Prices data for public use at `https://www.kaggle.com/datasets/thomasnibb/amsterdam-house-price-prediction`. In this chapter, we will use the dataset for the different recipes. The data is also available in the repository.

How to do it...

We will learn how to identify outliers using boxplots. We will use libraries such as `pandas`, `matplotlib`, and `seaborn`:

1. Import the relevant libraries:

```
import pandas as pd
import matplotlib.pyplot as plt
import seaborn as sns
```

2. Load the `.csv` into a dataframe using `read_csv` and subset the dataframe to include the relevant columns:

```
houseprice_data = pd.read_csv("data/HousingPricesData.csv")
houseprice_data = houseprice_data[['Price', 'Area', 'Room']]
```

3. Check the first five rows using the `head` method. Check the number of columns and rows:

```
houseprice_data.head()
      Price         Area Room
0    685000.0        64    3
1    475000.0        60    3
2    850000.0       109    4
3    580000.0       128    6
4    720000.0       138    5


houseprice_data.shape
(924, 3)
```

4. Get summary statistics on the dataframe variables using the `describe` method within the pandas library. Use a `lambda` function within the pandas `apply` method to suppress the scientific format:

```
houseprice_data.describe().apply(lambda s: s.apply('{0:.1f}'.
format))
             Price        Area       Room
count        920.0       924.0      924.0
mean      622065.4        96.0        3.6
std       538994.2        57.4        1.6
min       175000.0        21.0        1.0
25%       350000.0        60.8        3.0
50%       467000.0        83.0        3.0
75%       700000.0       113.0        4.0
max      5950000.0       623.0       14.0
```

5. Create boxplots to identify outliers across the three variables using the `boxplot` function within `seaborn`:

```
fig, ax = plt.subplots(1, 3,figsize=(40,18))

sns.set(font_scale = 3)
ax1 = sns.boxplot(data= houseprice_data, x= 'Price', ax = ax[0])
ax1.set_xlabel('Price in millions', fontsize = 30)
ax1.set_title('Outlier Analysis of House prices', fontsize = 40)


ax2 = sns.boxplot(data= houseprice_data, x= 'Area', ax=ax[1])
ax2.set_xlabel('Area', fontsize = 30)
```

```
ax2.set_title('Outlier Analysis of Areas', fontsize = 40)

ax3 = sns.boxplot(data= houseprice_data, x= 'Room', ax=ax[2])
ax3.set_xlabel('Rooms', fontsize = 30)
ax3.set_title('Outlier Analysis of Rooms', fontsize = 40)
plt.show()
```

This results in the following output:

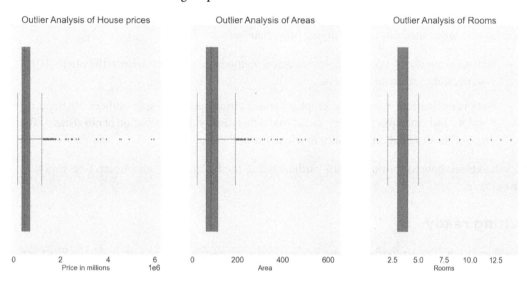

Figure 9.1: Outlier analysis on house prices, area, and rooms

Awesome. We just performed an outlier analysis on our dataset.

How it works...

In this recipe, we use the pandas, matplotlib, and seaborn libraries. In *step 2*, we load the house price data using the read_csv function and subset the data to select only relevant columns. In *step 3*, we examine the first five rows using the head method and check the shape (number of rows and columns) using the shape attribute.

In *step 4*, we generate summary statistics using the describe method in pandas. We use the format function in a lambda function to suppress the scientific notation and display the figures to 1 decimal place. The summary statistics give us a sense of the data distribution and provide pointers to the presence of outliers in the dataset. For example, we can see a wide gap between the minimum and maximum values. We also see a wide gap between the mean and maximum values. This wide gap is a pointer to the presence of outliers, in this case, outliers with very high values.

In *step 5*, we use the `boxplot` function in `seaborn` to identify the outliers. Outliers are the values beyond the upper whisker limit and below the lower whisker limit. They are represented as black circles outside the whisker limits.

Spotting univariate outliers

Univariate outliers are very large or small values that occur in a single variable in our dataset. These values are considered to be extreme and are usually different from the rest of the values in the variable. It is important to identify them and deal with them before any further analysis or modeling is done.

There are two major methods for identifying univariate outliers:

- **Statistical measures**: We can employ statistical methods such as the **interquartile range (IQR)**, Z-score, and measure of skewness.

- **Data visualization**: We can also employ various visual options to spot outliers. Histograms, boxplots, and violin plots are very useful charts that display the distribution of our dataset. The shape of the distribution can point to where the outliers lie.

We will explore how to spot univariate outliers using the `histplot` and `boxplot` function in `seaborn`.

Getting ready

We will work with the Amsterdam House Prices data for this recipe. You can retrieve all the files from the GitHub repository.

How to do it...

We will learn how to spot univariate outliers using libraries such as `pandas`, `matplotlib`, and `seaborn`:

1. Import the relevant libraries:

    ```
    import pandas as pd
    import matplotlib.pyplot as plt
    import seaborn as sns
    ```

2. Load the `.csv` into a dataframe using `read_csv` and subset the dataframe to include the relevant columns:

    ```
    houseprice_data = pd.read_csv("data/HousingPricesData.csv")
    houseprice_data = houseprice_data[['Price', 'Area', 'Room']]
    ```

3. Check the first five rows using the head method. Check the number of columns and rows:

```
houseprice_data.head()
      Price        Area    Room
0     685000.0     64      3
1     475000.0     60      3
2     850000.0     109     4
3     580000.0     128     6
4     720000.0     138     5


houseprice_data.shape
(924, 3)
```

4. Get summary statistics on the dataframe variables using the describe method within the pandas library. Use a lambda function within the pandas apply method to suppress the scientific format:

```
houseprice_data.describe().apply(lambda s: s.apply('{0:.1f}'.
format))
           Price       Area       Room
count      920.0       924.0      924.0
mean       622065.4    96.0       3.6
std        538994.2    57.4       1.6
min        175000.0    21.0       1.0
25%        350000.0    60.8       3.0
50%        467000.0    83.0       3.0
75%        700000.0    113.0      4.0
max        5950000.0   623.0      14.0
```

5. Calculate the IQR using the quantile method within pandas:

```
Q1 = houseprice_data['Price'].quantile(0.25)
Q3 = houseprice_data['Price'].quantile(0.75)
IQR = Q3 - Q1
print(IQR)
350000.0
```

6. Identify univariate outliers using IQR:

```
houseprice_data.loc[(houseprice_data['Price'] < (Q1 - 1.5 *
IQR)) | (houseprice_data['Price'] > (Q3 + 1.5 * IQR)),'Price']
20      1625000.0
28      1650000.0
31      1950000.0
33      3925000.0
57      1295000.0
```

```
                 . . .
885        1450000.0
902        1300000.0
906        1250000.0
910        1698000.0
917        1500000.0
```

7. Plot a histogram to spot univariate outliers using the `histplot` function in `seaborn`:

```
plt.figure(figsize = (40,18))

ax = sns.histplot(data= houseprice_data, x= 'Price')
ax.set_xlabel('Price in millions', fontsize = 30)
ax.set_ylabel('Count', fontsize = 30)
ax.set_title('Outlier Analysis of House prices', fontsize = 40)

plt.xticks(fontsize = 30)
plt.yticks(fontsize = 30)
```

This results in the following output:

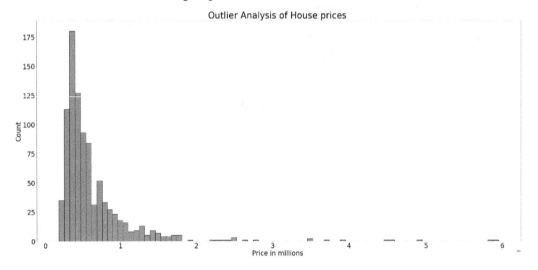

Figure 9.2: Univariate outlier analysis using a histogram

8. Plot a boxplot to spot univariate outliers using the `boxplot` function in `seaborn`:

```
plt.figure(figsize = (40,18))


ax = sns.boxplot(data= houseprice_data, x= 'Price')
ax.set_xlabel('Price in millions', fontsize = 30)
ax.set_title('Outlier Analysis of House prices', fontsize = 40)


plt.xticks(fontsize = 30)
plt.yticks(fontsize = 30)
```

This results in the following output:

Figure 9.3: Univariate outlier analysis using a boxplot

Great. We just spotted univariate outliers within our dataset.

How it works...

In *step 1*, we import the relevant libraries, such as `pandas`, `matplotlib`, and `seaborn`. In *step 2*, we load the house price data into a `pandas` dataframe using `read_csv` and select only the relevant columns.

In *steps 3* and *4*, we inspect the dataset using the `head` method and `shape` attribute. We then generate summary statistics using the `describe` method; the `format` function is used to suppress the scientific notation to display the numbers to 1 decimal place. The summary statistics give us a sense of the data distribution and provide pointers to the presence of outliers in the dataset.

In *steps 5* and *6*, we calculate the IQR using the `quantile` method in `pandas` and then apply the IQR formula for identifying univariate outliers to our dataset. The formula computes two thresholds, one to handle extremely low values and the other for extremely high values. For the former, we subtract 1.5 times the IQR from the first quartile (Q1), while for the latter, we add 1.5 times the IQR to the third quartile (Q3). If a value is below the former threshold, it is considered an outlier, and if it is beyond the latter threshold, it is also considered an outlier.

In *step 7*, we use the `histplot` function in `seaborn` to identify univariate outliers. Outliers, in this case, will be the isolated bars in the histogram. In our histogram, house prices beyond the 1.8 million mark can be considered outliers.

In *step 8*, we use the `boxplot` function in `seaborn` to identify the univariate outliers. Outliers are the values beyond the upper whisker limit. They are represented as the black circles outside the upper whisker limit. In the boxplot, the outliers seem to be values beyond the 1.2 million mark.

Finding bivariate outliers

Bivariate outliers are usually large or small values that occur in two variables simultaneously. In simple terms, these values differ from other observations when we examine the two variables together. Individually, the values in each variable may or may not be outliers; however, collectively, they are outliers.

To detect bivariate outliers, we typically need to check the relationship between the two variables. One primary method is to visualize the relationship using a scatter plot. Sometimes, we may be interested in identifying extreme values in a numerical variable across categories of a categorical variable or discrete values; in this case, a boxplot can be used. Using the boxplot, we can easily identify contextual outliers, which are usually observations considered anomalous given a specific context. The contextual outlier significantly deviates from the rest of the data points within a specific context. For example, when analyzing house prices, we can analyze attributes such as the location, number of bedrooms along with it. Contextual outliers are unlikely to appear as outliers when only univariate analysis is performed on house prices. They are often identified when we consider the house prices within a specific context. The characteristics or attributes of the houses such as location or number of rooms are examples of this specific context.

Again, it is important to understand the context and the objective of the analysis before choosing an appropriate method for handling bivariate outliers.

We will explore how to identify bivariate outliers using the `scatterplot` and `boxplot` methods in `seaborn`.

Getting ready

We will work with the Amsterdam House Prices data for this recipe. You can retrieve all the files from the GitHub repository.

How to do it...

We will learn how to identify bivariate outliers using scatterplots and boxplots. We will use libraries such as pandas, matplotlib, and seaborn:

1. Import the relevant libraries:

    ```
    import pandas as pd
    import matplotlib.pyplot as plt
    import seaborn as sns
    ```

2. Load the .csv into a dataframe using read_csv and subset the dataframe to include the relevant columns:

    ```
    houseprice_data = pd.read_csv("data/HousingPricesData.csv")
    houseprice_data = houseprice_data[['Price', 'Area', 'Room']]
    ```

3. Check the first five rows using the head method. Check the number of columns and rows:

    ```
    houseprice_data.head()
          Price      Area    Room
    0     685000.0     64       3
    1     475000.0     60       3
    2     850000.0    109       4
    3     580000.0    128       6
    4     720000.0    138       5


    houseprice_data.shape
    (924, 3)
    ```

4. Plot a scatterplot to identify bivariate outliers using the scatterplot function in seaborn:

    ```
    plt.figure(figsize = (40,18))


    ax = sns.scatterplot(data= houseprice_data, x= 'Price', y =
    'Area', s = 200)
    ax.set_xlabel('Price', fontsize = 30)
    ax.set_ylabel('Area',  fontsize = 30)
    plt.xticks(fontsize=30)
    plt.yticks(fontsize=30)


    ax.set_title('Bivariate Outlier Analysis of House prices and
    Area', fontsize = 40)
    ```

This results in the following output:

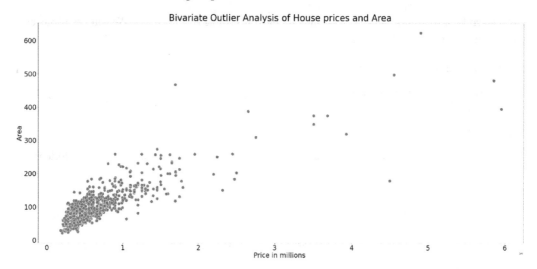

Figure 9.4: Bivariate outlier analysis using a scatterplot

5. Plot a boxplot to spot bivariate outliers using the `boxplot` function in `seaborn`:

```
plt.figure(figsize = (40,18))

ax = sns.boxplot(data= houseprice_data,x = 'Room', y= 'Price')
ax.set_xlabel('Room', fontsize = 30)
ax.set_ylabel('Price',  fontsize = 30)
plt.xticks(fontsize=30)
plt.yticks(fontsize=30)
ax.set_title('Bivariate Outlier Analysis of House prices and
Rooms', fontsize = 40)
```

This results in the following output:

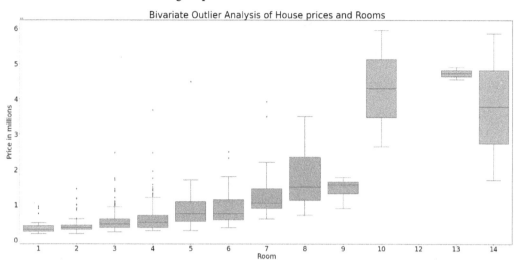

Figure 9.5: Univariate outlier analysis using a boxplot

Well done. We just identified bivariate outliers within our dataset.

How it works...

In *steps 1* and *2*, we import the relevant libraries and load our house price data from a `.csv` file into a dataframe using the `read_csv` method. We also select only relevant columns for the analysis.

In *step 3*, we inspect our data using the `head` and `shape` methods. In *step 4*, we use the `scatterplot` function in `seaborn` to identify bivariate outliers. Outliers are points in the scatterplot that seem separated from other points. In our scatterplot, house prices beyond the 1.8 million and 300 **square meter (sqm)** mark can be considered outliers since these data points are isolated.

In *step 5*, we use the `boxplot` function in `seaborn` to identify the bivariate outliers. Outliers can be identified within the categories as the values that are beyond the upper whisker limit in a specific category. Outliers that appear in 1–4-bedroom houses are likely to be contextual outliers. These outliers are unlikely to appear when we perform a univariate analysis of house prices since most of their values are below 2 million. However, they appear as outliers when we consider the context of house prices based on the number of bedrooms.

Identifying multivariate outliers

Multivariate outliers are extreme values that occur in three or more variables simultaneously. In simple terms, the observation differs from other observations when the variables are considered collectively and not necessarily individually.

We can identify outliers using techniques such as the following:

- **Mahalanobis distance**: It measures the distance of each observation from the center of the data distribution (centroid). In simple terms, Mahalanobis distance helps to quantify how far away each observation is from the mean vector (set of mean values of each variable in a dataset which acts as a centroid or central location of the variables in the dataset). It provides a measure of distance that is adjusted for the specific characteristics of the dataset such as the correlations between variables and the variability of each variable, enabling a more accurate assessment of similarity between observations. This allows for the identification of outliers that deviate significantly from the overall distribution (observations with a high distance value).

- **Cluster analysis**: It helps to group similar observations together and to find multivariate outliers. **Density-based spatial clustering of applications with noise (DBSCAN)** is a popular clustering algorithm that can be used for outlier detection. It identifies groups of points that are close together and places them in various clusters; it then considers points that are not part of any cluster as outliers It also considers points in small clusters as outliers.

- **Visualization**: We can visualize the output of the preceding methods to inspect the outliers.

We will explore how to identify multivariate outliers using the `malahanobis` function in `scipy`, the `DBSCAN` class in `scikit-learn`, and `numpy` functions such as `mean`, `cov`, and `inv`.

Getting ready

We will work with the Amsterdam House Prices data for this recipe. You can retrieve all the files from the GitHub repository.

How to do it...

We will learn how to spot multivariate outliers using libraries such as `numpy`, `pandas`, `matplotlib`, `seaborn`, `scikit-learn`, and `scipy`:

1. Import the relevant libraries:

```
import numpy as np
import pandas as pd
import matplotlib.pyplot as plt
import seaborn as sns
from sklearn.preprocessing import StandardScaler
```

```
from scipy.spatial.distance import mahalanobis
from sklearn.cluster import DBSCAN
from sklearn.neighbors import NearestNeighbors
```

2. Load the `.csv` into a dataframe using `read_csv` and subset the dataframe to include the relevant columns:

```
houseprice_data = pd.read_csv("data/HousingPricesData.csv")
houseprice_data = houseprice_data[['Price', 'Area', 'Room'
,'Lon','Lat']]
```

3. Check the first five rows using the `head` method. Check the number of columns and rows:

```
houseprice_data.head()
      Price       Area    Room    Lon        Lat
0     685000.0    64      3       4.907736   52.356157
1     475000.0    60      3       4.850476   52.348586
2     850000.0    109     4       4.944774   52.343782
3     580000.0    128     6       4.789928   52.343712
4     720000.0    138     5       4.902503   52.410538
```

```
houseprice_data.shape
(924, 5)
```

4. Scale the data to ensure all variables have a similar scale using the `StandardScaler` class in `scikit-learn`:

```
scaler = StandardScaler()

houseprice_data_scaled = scaler.fit_transform(houseprice_data)
houseprice_data_scaled = pd.DataFrame(houseprice_data_scaled,
columns=houseprice_data.columns)
houseprice_data_scaled.head()
      Price     Area     Room      Lon       Lat
0     0.1168   -0.5565   -0.35905   0.3601   -0.2985
1    -0.2730   -0.626    -0.3590   -0.7179   -0.6137
2     0.4231    0.2272    0.2692    1.0575   -0.8138
3    -0.0780    0.5581    1.5259   -1.8579   -0.8167
4     0.1817    0.7323    0.8976    0.2616    1.9659
```

5. Calculate the Mahalanobis distance of the variables using the following methods and functions: `mean` and `cov` in pandas, `inv` in numpy, and `mahalanobis` in scipy:

```
mean = houseprice_data_scaled.mean()
cov = houseprice_data_scaled.cov()
inv_cov = np.linalg.inv(cov)

distances = []
for _, x in houseprice_data_scaled.iterrows():
    d = mahalanobis(x, mean, inv_cov)
    distances.append(d)

houseprice_data['mahalanobis_distances'] = distances
houseprice_data
```

This results in the following output:

	Price	Area	Room	Lon	Lat	mahalanobis_distances
0	685000.0	64	3	4.907736	52.356157	1.322730
1	475000.0	60	3	4.850476	52.348586	1.312112
2	850000.0	109	4	4.944774	52.343782	1.380465
3	580000.0	128	6	4.789928	52.343712	2.877517
4	720000.0	138	5	4.902503	52.410538	2.416231
...	...	...	...	...	...	...
919	750000.0	117	1	4.927757	52.354173	3.427566
920	350000.0	72	3	4.890612	52.414587	2.272896
921	350000.0	51	3	4.856935	52.363256	1.115096
922	599000.0	113	4	4.965731	52.375268	1.749725
923	300000.0	79	4	4.810678	52.355493	1.840197

Figure 9.6: Mahalanobis distances of a multivariate dataset

6. Plot the Mahalanobis distances in a boxplot to identify multivariate outliers:

```
plt.figure(figsize = (40,18))


ax = sns.boxplot(data= houseprice_data, x= 'mahalanobis_
distances')
ax.set_xlabel('Mahalanobis Distances',  fontsize = 30)
plt.xticks(fontsize=30)
ax.set_title('Outlier Analysis of Mahalanobis Distances',
fontsize = 40)
```

This results in the following output:

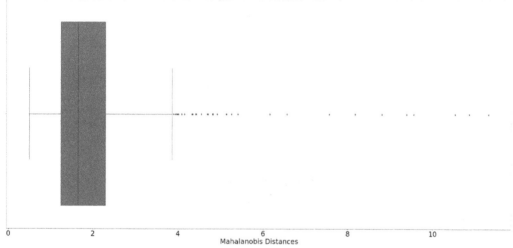

Figure 9.7: Boxplot of Mahalanobis distances

7. Identify multivariate outliers by setting a threshold based on the limits in the boxplot:

```
mahalanobis_outliers = houseprice_data[houseprice_
data['mahalanobis_distances']>=4]
mahalanobis_outliers.head()
```

This results in the following output:

	Price	Area	Room	Lon	Lat	mahalanobis_distances
31	1950000.0	258	4	4.887444	52.385346	4.547578
33	3925000.0	319	7	4.875471	52.361571	6.566091
103	4550000.0	497	13	4.898620	52.358798	7.558049
156	799000.0	230	8	4.961381	52.389087	4.346119
179	4495000.0	178	5	4.894290	52.373106	10.876237

Figure 9.8: Outliers based on the Mahalanobis distance

8. Remove missing values using the `dropna` method in `pandas` in preparation for DBSCAN:

```
houseprice_data_no_missing = houseprice_data_scaled.dropna()
```

9. Create a k-distance graph to identify optimal parameters for DBSCAN:

```
knn = NearestNeighbors(n_neighbors=2)
nbrs = knn.fit(houseprice_data_no_missing)
distances, indices = nbrs.kneighbors(houseprice_data_no_missing)

distances = np.sort(distances, axis=0)
distances = distances[:,1]
plt.figure(figsize=(40,18))
plt.plot(distances)
plt.title('K-distance Graph',fontsize=40)
plt.xlabel('Data Points sorted by distance',fontsize=30)
plt.ylabel('Epsilon',fontsize=30)
plt.xticks(fontsize=30)
plt.yticks(fontsize=30)
plt.show()
```

This results in the following output:

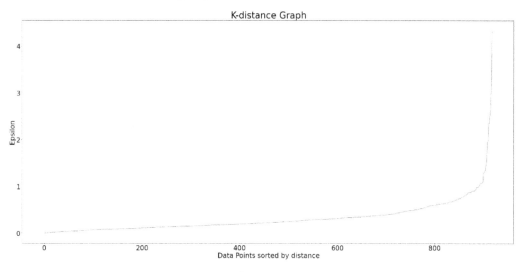

Figure 9.9: K-distance graph to identify the optimal Epsilon value

10. Identify multivariate outliers using the DBSCAN class in `scikit-learn`:

```
dbscan = DBSCAN(eps= 1, min_samples=6)
dbscan.fit(houseprice_data_no_missing)
labels = dbscan.labels_
n_clusters = len(set(labels))
outlier_indices = np.where(labels == -1)[0]

print(f"Number of clusters: {n_clusters}")
print(f"Number of outliers: {len(outlier_indices)}")
```

11. View multivariate outliers:

```
houseprice_data.loc[outlier_indices,:].head()
```

This results in the following output:

	Price	Area	Room	Lon	Lat	mahalanobis_distances
28	1650000.0	235	7	4.820848	52.358631	2.802948
31	1950000.0	258	4	4.887444	52.385346	4.547578
33	3925000.0	319	7	4.875471	52.361571	6.566091
87	995000.0	97	2	4.884982	52.369202	1.956290
102	725000.0	95	3	4.886006	52.377377	0.842156

Figure 9.10: Outliers from the DBSCAN outlier detection

Good job. Now we have identified multivariate outliers in our dataset.

How it works...

In steps 1 to 2, we import the relevant libraries and load our house price data from a .csv file into a dataframe. We also subset the data to include only relevant columns.

In steps 3 to 4, we get a glimpse of the first 5 rows of our data using the head and we check the shape using the shape method. We then scale our dataset using the StandardScaler class in scikit-learn. This is a prerequisite for Mahalanobis distance and DBSCAN because they are distance-based algorithms.

In step 5, we calculate the mean, covariance and inverse of the covariance matrix of our dataset as these are parameters for Mahalanobis outlier detection. We use the mean, and cov methods in pandas and inv in numpy. We then calculate the Mahalanobis distance of our dataset using the mahalanobis function in scipy and save the distances first in a list and then in a dataframe column. The covariance and inverse of covariance help us to adjust for the specific characteristics of the dataset such as the correlations between variables and the variability of each variable so that the assessment of similarity or dissimilarity between observations is more accurate.

In steps 6 to 7, we use the boxplot method in seaborn to identify the outlier Mahalanobis distances. From the boxplot, distances beyond the 4 mark are outliers. We then apply this threshold to our dataset to generate the multivariate outliers.

In steps 8 to 9, we remove missing values because this is a requirement for the DBSCAN algorithm There are two key parameters which the DBSCAN algorithm is sensitive to. The parameters are `epsilon` and `minPoints`. DBSCAN uses the `minPoints` parameter to determine the minimum number of neighboring points that should form a cluster while it uses `epsilon` parameter to determine the maximum distance between points for them to be considered neighboring points.

To identify the optimal epsilon value, we create a k-distance graph. For the graph, we start by finding the distance between each data point in the dataset and its nearest neighbor using the `NearestNeighbors` class from `scikit-learn`. We use the `fit` method to fit the nearest neighbor model to our dataset and then we apply the `kneighbors` method to extract the distance between the nearest neighbor of each point and the corresponding indices of each point and its nearest neighbor. We sort the distances in ascending order and in the distance array, we select the values in the second column of distances. This column contains the distances of each data point to its nearest neighbor while the first column corresponds to the distance of each data point to itself. We then use the `plot` method to plot the distances. The plot displays distance on the y-axis while by default it displays the indices of the array on the x-axis (0 to 920). The optimal epsilon is found at the elbow of the chart. For the optimal `minPoints`, we can select a value that is at least 1 point greater than the number of variables.

In step 10, we create a DBSCAN model using the DBSCAN class in `scikit-learn`. We use `min_samples` (minPoints) of 6 and an `epsilon` of 1. The epsilon value is derived from the k-distance graph. We use the value at the elbow of the graph which is approximately 1. We extract the labels using the `labels_` attribute of the class and we extract the number of clusters by performing a distinct count on the labels using the `len` and `set` functions. We then save the indices of the outliers using the `where` method in `numpy`. Outliers in DBSCAN are typically labelled as -1.

In *step 11*, we view the outliers. Based on the output, we notice a difference between the outliers generated by the Mahalanobis distance and DBSCAN. This can be attributed to the fact that the algorithms use different approaches for outlier detection. Hence, it is important to leverage domain knowledge to identify which algorithm provides the best results.

See also

You can check out this useful resource:

- `https://www.analyticsvidhya.com/blog/2020/09/how-dbscan-clustering-works/`

Flooring and capping outliers

Quantile-based flooring and capping are two related outlier handling techniques. They involve replacing extreme values with fixed values, in this case, quantiles.

Flooring involves replacing small extreme values with a predetermined minimum value, such as the value of the 10th percentile. On the other hand, capping involves replacing large extreme values with a predetermined maximum value, such as the value of the 90th percentile.

These techniques are more appropriate when extreme values are likely caused by measurement errors or data entry errors. In cases where the outliers are genuine, these techniques will likely introduce bias.

We will explore how to handle outliers using the flooring and capping approach. We will use the `quantile` method in `pandas` to achieve this.

Getting ready

We will work with the Amsterdam House Prices data for this recipe. You can retrieve all the files from the GitHub repository.

How to do it...

We will learn how to handle outliers using the flooring and capping approach. We will use libraries such as `numpy`, `pandas`, `matplotlib`, and `seaborn`:

1. Import the relevant libraries:

    ```
    import numpy as np
    import pandas as pd
    import matplotlib.pyplot as plt
    import seaborn as sns
    ```

2. Load the `.csv` into a dataframe using `read_csv` and subset the dataframe to include the relevant columns:

    ```
    houseprice_data = pd.read_csv("data/HousingPricesData.csv")
    houseprice_data = houseprice_data[['Price', 'Area', 'Room']]
    ```

3. Check the first five rows using the `head` method. Check the number of columns and rows:

    ```
    houseprice_data.head()
         Price       Area    Room
    0    685000.0    64      3
    1    475000.0    60      3
    ```

```
2     850000.0    109    4
3     580000.0    128    6
4     720000.0    138    5

houseprice_data.shape
(924, 3)
```

4. Identify quantile thresholds using the `quantile` method in `pandas`:

```
floor_thresh = houseprice_data['Price'].quantile(0.10)
cap_thresh = houseprice_data['Price'].quantile(0.90)
print(floor_thresh,cap_thresh)
285000 1099100
```

5. Perform flooring and capping using quantile thresholds:

```
houseprice_data['Adjusted_Price'] = houseprice_data.
loc[:,'Price']
houseprice_data.loc[houseprice_data['Price'] < floor_thresh,
'Adjusted_Price'] = floor_thresh
houseprice_data.loc[houseprice_data['Price'] > cap_thresh,
'Adjusted_Price'] = cap_thresh
```

6. Visualize the floored and capped dataset in a histogram using `histplot` in `seaborn`:

```
plt.figure(figsize = (40,18))


ax = sns.histplot(data= houseprice_data, x= 'Adjusted_Price')
ax.set_xlabel('Adjusted Price in millions',fontsize = 30)
ax.set_title('Outlier Analysis of Adjusted House prices',
fontsize = 40)
ax.set_ylabel('Count',  fontsize = 30)
plt.xticks(fontsize=30)
plt.yticks(fontsize=30)
```

This results in the following output:

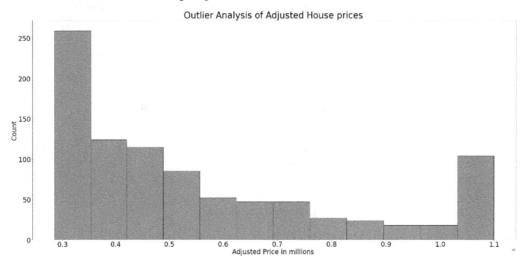

Figure 9.11: Histogram displaying floored and capped dataset

7. Identify group flooring and capping thresholds using the `nanpercentile` function in `numpy` and the `transform` method in `pandas`. This will help perform flooring and capping on contextual outliers:

```
houseprice_data['90_perc'] = houseprice_data.groupby('Room')
['Price'].transform(lambda x: np.nanpercentile(x,90))
houseprice_data['10_perc'] = houseprice_data.groupby('Room')
['Price'].transform(lambda x: np.nanpercentile(x,10))
```

8. Perform flooring and capping using the group quantile thresholds:

```
houseprice_data['Group_Adjusted_Price'] = houseprice_data.
loc[:,'Price']
houseprice_data.loc[houseprice_data['Price'] < houseprice_
data['10_perc'], 'Group_Adjusted_Price'] = houseprice_data['10_
perc']
houseprice_data.loc[houseprice_data['Price'] > houseprice_
data['90_perc'], 'Group_Adjusted_Price'] = houseprice_data['90_
perc']
```

9. Visualize the floored and capped dataset in a histogram using `boxplot` in `seaborn`:

```
plt.figure(figsize = (40,18))

ax = sns.boxplot(data= houseprice_data,x = 'Room', y= 'Group_
Adjusted_Price')
ax.set_xlabel('Room', fontsize = 30)
```

```
ax.set_ylabel('Adjusted Price in millions',  fontsize = 30)
plt.xticks(fontsize=30)
plt.yticks(fontsize=30)
ax.set_title('Bivariate Outlier Analysis of Adjusted House
prices and Rooms', fontsize = 40)
```

This results in the following output:

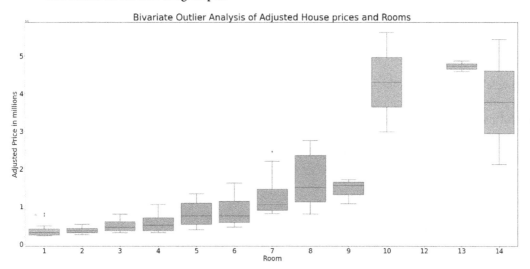

Figure 9.12: Boxplot displaying floored and capped dataset

Good. Now we have floored and capped our dataset using quantile thresholds.

How it works...

In *step 1*, we import the numpy, pandas, matplotlib, and seaborn libraries. In *step 2*, we load the house price data into a dataframe. In *steps 3* and *4*, we examine the first five rows using the head method and check the shape (number of rows and columns) using the shape attribute. We also create the quantile thresholds for flooring and capping using the quantile method in pandas.

In *steps 5* and *6*, we apply the thresholds to the dataset. We create a new price column and we replace any value in the price column that is below the flooring threshold value with the flooring threshold value. We also replace any value above the capping threshold value with capping threshold value. Then we use the histplot function in seaborn to display the new distribution of the floored and capped dataset. A key observation from the histogram is the absence of the isolated bars, which were outliers.

In *step 7* and *step 8*, we perform group flooring and capping, which is ideal for identifying contextual outliers. Even though some 1- or 2-bedroom house prices may not appear as outliers when analyzed globally, they appear as outliers when we compare them to house prices within the same category, that is when we analyze them locally. These house prices are contextual outliers because they are outliers within the context of the number of rooms a house has. First, we identify the group thresholds using the `groupby` and `transform` methods in `pandas` and the `nanpercentile` function in `numpy`. This approach helps us calculate the quantile thresholds within each number-of-rooms category (1, 2, and 3 bedrooms, and so on). We then apply these thresholds to the dataset.

In *step 9*, we use the `boxplot` function in `seaborn` to display the new distribution of the floored and capped dataset. A key observation from the boxplot is the fact that most contextual outliers in the categories have been handled and cannot be spotted.

Removing outliers

A simple approach to handling outliers is to remove them completely before analyzing our dataset; this is also known as trimming. A major setback of this approach is the fact that we may lose some useful insights, especially if the outliers were legitimate. Therefore, it is very important to understand the context of the dataset before removing outliers. In certain scenarios, edge cases exist, and these cases can easily be tagged as outliers when the context isn't properly understood. Edge cases are typically scenarios that are unlikely to occur. However, they can reveal important insights that will be overlooked if they are removed.

Trimming can be useful when the distribution of the data is important and we need to retain it. It is also useful when we have a minimal number of outliers.

We will explore how to remove outliers from our dataset using the `drop` method in `pandas` to achieve this.

Getting ready

We will work with the Amsterdam House Prices data for this recipe. You can retrieve all the files from the GitHub repository.

How to do it...

We will learn how to remove outliers and visualize the output using libraries such as `pandas`, `matplotlib`, and `seaborn`:

1. Import the relevant libraries:

```
import pandas as pd
import matplotlib.pyplot as plt
import seaborn as sns
```

2. Load the .csv into a dataframe using `read_csv` and subset the dataframe to include the relevant columns:

```
houseprice_data = pd.read_csv("data/HousingPricesData.csv")
houseprice_data = houseprice_data[['Price', 'Area', 'Room']]
```

3. Check the first five rows using the `head` method. Check the number of columns and rows:

```
houseprice_data.head()
     Price       Area    Room
0    685000.0    64      3
1    475000.0    60      3
2    850000.0    109     4
3    580000.0    128     6
4    720000.0    138     5

houseprice_data.shape
(924, 3)
```

4. Calculate the IQR using the `quantile` method in pandas:

```
Q1 = houseprice_data['Price'].quantile(0.25)
Q3 = houseprice_data['Price'].quantile(0.75)
IQR = Q3 - Q1
print(IQR)
350000
```

5. Identify outliers using the IQR and save them in a dataframe:

```
outliers = houseprice_data['Price'][(houseprice_data['Price']
< (Q1 - 1.5 * IQR)) |(houseprice_data['Price'] > (Q3 + 1.5 *
IQR))]

outliers.shape
(71,)
```

6. Remove outliers from the dataset using the `drop` method in pandas:

```
houseprice_data_no_outliers = houseprice_data.drop(outliers.
index, axis = 0)

houseprice_data_no_outliers.shape
(853,3)
```

7.   Visualize the dataset without outliers using the `histplot` method in seaborn:

```
plt.figure(figsize = (40,18))

ax = sns.histplot(data= houseprice_data_no_outliers, x= 'Price')
ax.set_xlabel('Price in millions', fontsize = 30)
ax.set_title('House prices with no Outliers', fontsize = 40)
ax.set_ylabel('Count',  fontsize = 30)
plt.xticks(fontsize=30)
plt.yticks(fontsize=30)
```

This results in the following output:

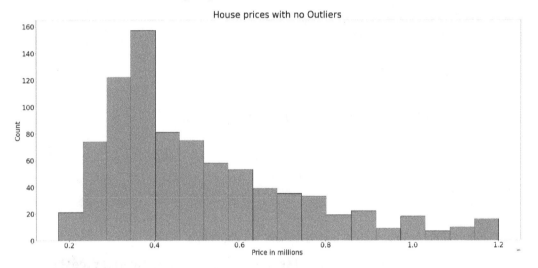

Figure 9.13: Histogram displaying the dataset with outliers removed

Nice. We have removed outliers from our dataset.

How it works...

In *steps 1* and *2*, we import the relevant libraries and load the house price data into a dataframe using the `read_csv` function in `pandas`. We also subset to include only relevant columns. In *steps 3* to *5*, we get a glimpse of the first five rows using the `head` method and check the shape of the data using the `shape` attribute. We then calculate the IQR using the `quantile` method in `pandas` and apply the IQR formula to extract the outliers in our dataset. The formula computes two thresholds, one to handle extremely low values and the other for extremely high values. The threshold for low values is equal to the first quartile (`Q1`) minus `1.5` times the IQR, while the threshold for high values equals the third quartile plus `1.5` times the IQR. We save the outliers in a dataframe.

In *step 6*, we remove the outliers from our dataset by using the drop method in pandas and passing the outlier indices as the first argument and the axis value of 0 as the second argument. An axis value of 0 refers to dropping rows, while a value of 1 refers to dropping columns.

In *step 7*, we use the histplot function in seaborn to create a histogram of our dataset without the outliers. A key observation is the absence of the isolated bars, which represented outliers.

Replacing outliers

Another approach that can be considered to handle outliers is replacing the extreme values with a predetermined value. Just like the removal of outliers, this needs to be done with utmost care because it can introduce bias into our dataset. Flooring and capping are also forms of replacing outliers. However, in this recipe, we will focus on other methods:

- **Statistical measures**: This involves replacing outliers with the mean, median, or percentiles of the dataset

- **Interpolation**: This involves estimating the value of an outlier using the neighboring data points of the outlier

- **Model-based methods**: These involve using a machine learning model to predict the replacement value for the outliers

It is important to note that the preceding methods will affect the shape and characteristics of the dataset distribution, and they are not appropriate in scenarios where the distribution of the data is important.

We will explore how to replace outliers using the quantile and interpolation methods in pandas.

Getting ready

We will work with the Amsterdam House Prices data for this recipe. You can retrieve all the files from the GitHub repository.

How to do it...

We will learn how to remove outliers and visualize the output using libraries such as numpy, pandas, matplotlib, and seaborn:

1. Import the relevant libraries:

```
import numpy as np
import pandas as pd
import matplotlib.pyplot as plt
import seaborn as sns
```

2. Load the `.csv` into a dataframe using `read_csv` and subset the dataframe to include the relevant columns:

```
houseprice_data = pd.read_csv("data/HousingPricesData.csv")
houseprice_data = houseprice_data[['Price', 'Area', 'Room']]
```

3. Check the first five rows using the `head` method. Check the number of columns and rows:

```
houseprice_data.head()
      Price       Area     Room
0     685000.0     64        3
1     475000.0     60        3
2     850000.0    109        4
3     580000.0    128        6
4     720000.0    138        5

houseprice_data.shape
(924, 3)
```

4. Calculate the IQR using the `quantile` method in `pandas`:

```
Q1 = houseprice_data['Price'].quantile(0.25)
Q3 = houseprice_data['Price'].quantile(0.75)
IQR = Q3 - Q1
print(IQR)
350000
```

5. Identify replacements using quantiles:

```
low_replace = houseprice_data['Price'].quantile(0.05)
high_replace = houseprice_data['Price'].quantile(0.95)
median_replace = houseprice_data['Price'].quantile(0.5)
print(low_replace,high_replace,median_replace)
250000 1450000 467000
```

6. Replace outliers using the `quantile` method in `pandas`:

```
houseprice_data['Adjusted_Price'] = houseprice_data.
loc[:,'Price']
houseprice_data.loc[houseprice_data['Price'] < (Q1 - 1.5 *
IQR),'Adjusted_Price'] = low_replace
houseprice_data.loc[houseprice_data['Price'] > (Q3 + 1.5 * IQR),
'Adjusted_Price'] = high_replace
```

7. Identify outliers and replace them with `nan` in preparation for interpolation using the `nan` attribute in `numpy`:

```
outliers = houseprice_data['Price'][(houseprice_data['Price']
< (Q1 - 1.5 * IQR)) | (houseprice_data['Price'] > (Q3 + 1.5 *
IQR))]

houseprice_replaced_data = houseprice_data.copy()
houseprice_replaced_data.loc[:,'Price_with_nan'] = houseprice_
replaced_data.loc[:,'Price']
houseprice_replaced_data.loc[outliers.index,'Price_with_nan'] =
np.nan

houseprice_replaced_data.isnull().sum()
Price                4
Area                 0
Room                 0
Adjusted_Price       4
Price_with_nan      75
```

8. Use the `interpolate` function to replace outliers:

```
houseprice_replaced_data = houseprice_replaced_data.
interpolate(method='linear')
houseprice_replaced_data.loc[outliers.index,:].head()
```

This results in the following output:

	Price	Area	Room	Adjusted_Price	Price_with_nan
20	1625000.0	199	6	1450000.0	475000.0
28	1650000.0	235	7	1450000.0	350000.0
31	1950000.0	258	4	1450000.0	767500.0
33	3925000.0	319	7	1450000.0	605000.0
57	1295000.0	145	5	1450000.0	480000.0

Figure 9.14: Outliers values replaced with interpolation values

9. Visualize the interpolated dataset using the `histplot` function in `seaborn`:

```
plt.figure(figsize = (40,18))

ax = sns.histplot(data= houseprice_replaced_data, x= 'Price_
with_nan')
ax.set_xlabel('Interpolated House Prices in millions', fontsize
= 30)
```

```
ax.set_title('House prices with replaced Outliers', fontsize =
40)
ax.set_ylabel('Count',  fontsize = 30)
plt.xticks(fontsize=30)
plt.yticks(fontsize=30)
```

This results in the following output:

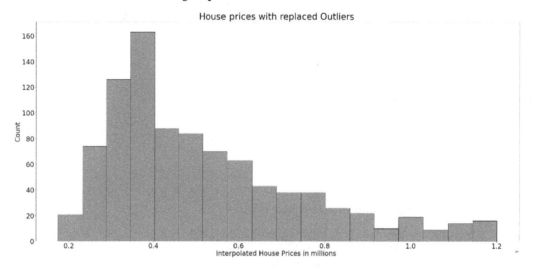

Figure 9.15: Histogram displaying interpolation values of outliers

Well done. We have replaced outliers using predetermined replacement values and interpolation.

How it works...

In *steps 1* and *2*, we import the relevant libraries and load the house price data into a dataframe using the read_csv method. We also select the relevant columns.

In *steps 3* to *5*, we inspect our data using the head and shape methods. We then compute the IQR using the quantile method in pandas. We also create replacement values using the quantile method. With this method, we can create replacement values based on high-valued quantiles, low-valued quantiles, or the median. In *step 6*, we replace the outliers in the dataset with the replacement values.

In *steps 7* and *8*, we replace outliers with a NaN value using the numpy nan attribute. This is a requirement for the interpolation outlier replacement method. We then use the interpolate method in pandas to estimate the values to replace outliers. In the method parameter, we use the linear interpolation technique; this fills in the values using a linear relationship between known values. Some other interpolation techniques include polynomial, cubic and more. These can be used when the data has a curvy trend.

In *step 9*, we use the histplot function in seaborn to visualize the interpolated dataset.

Identifying missing values

A missing value refers to the absence of a specific value within a variable. In structured data, it represents an empty cell in a dataframe and is sometimes represented as NA, NaN, NULL, and so on.

Missing values can lead to inaccurate conclusions and biased analysis; therefore, it is important to handle them when encountered in our dataset.

The following example illustrates this:

Class A	Class B	Class C
70	60	90
80	50	50
60	70	80
75	75	
	90	75

Class A	Class B	Class C
70	60	90
80	50	50
60	70	80
75	75	15
95	90	75

Figure 9.16: Class assessment scores with missing values (left) and without missing values (right)

From the preceding example, we can deduce the following:

- **With missing values (Left table):** **Class B** and **Class C** have on distinction (>=90) each, while **Class A** has none. The average scores across the classes are 71.25, 63.75, and 75, respectively. Class C has the highest average score.

- **Without missing values (Right table):** All classes have one distinction (>=90) each. The average scores across the classes are 76, 63.75, and 63, respectively. Class A has the highest average score, while Class C now has the lowest average score.

Understanding why missing values exist in a dataset is crucial to choosing the appropriate method to handle them. Some common mechanisms that result in missing data include the following:

- **Missing completely at random (MCAR):** In MCAR, the probability of data being missing is completely random and unrelated to other variables in the dataset. In simple terms, the data is missing due to chance, and there is no clear pattern to the missingness. In this case, the probability of an observation being missing is the same for all observations. When dealing with MCAR, it is safe to ignore the missing values during analysis. An example of MCAR is when participants mistakenly skip some fields randomly or when there is a random system glitch.

- **Missing at random (MAR):** In MAR, the probability of data being missing is related to other variables in the dataset. In simple terms, the missingness can be explained by other variables in the dataset, and once those variables are accounted for, the missing data is considered random. When dealing with MCAR, it is safe to use imputation methods to handle the missing values during analysis. An example of MAR is when some blue-collar workers refuse to provide their income when filling surveys which ask about income because they think it is very low. In the same vein, some white collar workers may refuse to provide their income in a survey because they think it is very high. When analyzing such a survey, the missing values in the income variable will be related to the job category variable or industry variable in the survey.

- **Missing not at random (MNAR)**: In MNAR, the probability of data being missing cannot be explained by other variables in the dataset but is related to some unobserved values. It points to the likelihood of a systematic pattern in the missing data. Handling this type of missingness can be very challenging. An example of MNAR typically happens in polls when a subset of people is completely excluded due to biased sampling mechanisms.

We will explore how to identify missing values using the `isnull` method in `pandas` and the `bar` and `matrix` methods in `missingno`.

To install `missingno`, use the following command:

```
pip install missingno
```

Getting ready

We will work with the Amsterdam House Prices data for this recipe. You can retrieve all the files from the GitHub repository.

How to do it...

We will learn how to remove outliers and visualize the output using libraries such as `pandas`, `matplotlib`, and `seaborn`:

1. Import the relevant libraries:

    ```
    import pandas as pd
    import matplotlib.pyplot as plt
    import seaborn as sns
    import missingno as msno
    ```

2. Load the `.csv` into a dataframe using `read_csv` and subset the dataframe to include the relevant columns:

    ```
    houseprice_data = pd.read_csv("data/HousingPricesData.csv")
    houseprice_data = houseprice_data[['Price', 'Area', 'Room']]
    ```

3. Check the first five rows using the head method. Check the number of columns and rows:

```
houseprice_data.head()
      Price       Area     Room
0    685000.0       64        3
1    475000.0       60        3
2    850000.0      109        4
3    580000.0      128        6
4    720000.0      138        5

houseprice_data.shape
(924, 3)
```

4. Identify missing values using the isnull attribute in pandas:

```
houseprice_data.isnull().sum()
Price    4
Area     0
Room     0
```

5. Identify missing values using the info method in pandas:

```
houseprice_data.info()
<class 'pandas.core.frame.DataFrame'>
RangeIndex: 924 entries, 0 to 923
Data columns (total 3 columns):
 #    Column   Non-Null Count   Dtype
---   ------   --------------   -----
 0    Price    920 non-null     float64
 1    Area     924 non-null     int64
 2    Room     924 non-null     int64
dtypes: float64(1), int64(2)
memory usage: 21.8 KB
```

6. Identify missing values using the bar method in missingno:

```
msno.bar(houseprice_data)
```

This results in the following output:

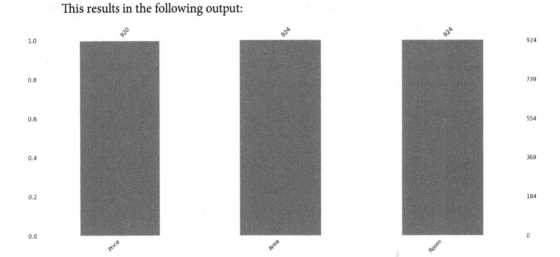

Figure 9.17: Bar chart displaying the count of complete values within each variable

7. Identify missing values using the `matrix` method in `missingno`:

```
msno.matrix(houseprice_data)
```

This results in the following output:

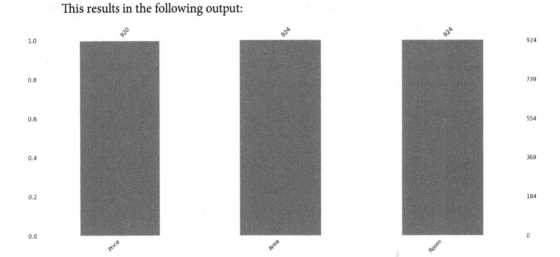

Figure 9.18: Matrix displaying the complete and missing values within each variable

8. View missing values using the `isnull` method in `pandas`:

```
houseprice_data[houseprice_data['Price'].isnull()]
         Price      Area      Room
73                  147         3
321                 366        12
610                 107         3
727                  81         3
```

Well done. Now we have identified missing values.

How it works...

In *steps 1* and *2*, we import the `pandas`, `matplotlib`, `seaborn`, and `missingno` libraries. We then load the house price data into a dataframe using the `read_csv` method and subset for relevant columns.

In *step 3*, we inspect the dataset using the `head` and `shape` methods. In *step 4*, we identify columns with missing values and the number of missing values using the `isnull` and `sum` methods in `pandas`. In *step 5*, we identify columns with missing values using the `info` attribute in `pandas`. This method displays the number of complete records per column.

In *steps 6* and *7*, we use the `bar` method in the `missingno` library to identify missing values across the columns in our dataset. We also use the `matrix` method in the `missingno` library to identify missing values across all columns in our dataset. The bar chart displays the number of complete records per column, while the matrix displays the number of missing values using horizontal lines drawn across the specific column with missing values.

In *step 8*, we subset our dataset to view the missing values using the isnull method in pandas.

Dropping missing values

A simple approach to handling missing values is to remove them completely. Some common approaches to removing missing values include listwise deletion and pairwise deletion. Listwise deletion involves removing all observations that contain one or more missing values. This approach is also known as complete-case analysis, meaning only complete cases are analyzed. On the other hand, in pairwise deletion, only the available data for each variable is used for analysis. Observations with missing values are included in the analysis. For each observation, the missing value is skipped/excluded for variables where the missing values exist. On the other hand, for variables without the missing value, the values present are used for analysis. Pairwise deletion is also known as available case analysis.

Listwise deletion leads to loss of data, while pairwise retains data. Both methods can sometimes introduce bias into the dataset. The following diagram explains the difference between listwise and pairwise deletion:

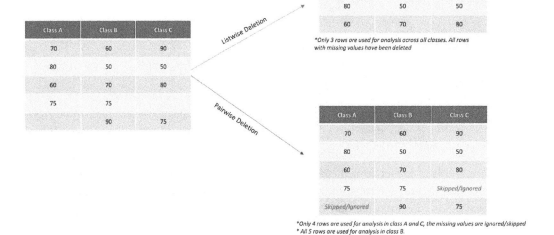

Figure 9.19: Listwise versus pairwise deletion

From the preceding example, we see that listwise deletion results in the analysis of only three rows across the classes. However, with pairwise deletion, the number of rows depends on the available cases per class. For Class A and Class C, only four rows with available data are used, while for Class B, all five rows are used because there are no missing values.

Another approach that can be used for removing missing values is the removal of entire columns where missing values exist. This approach is typically considered when the entire column is filled with missing values or most values in the column are missing.

We will explore how to drop missing values using the `dropna` method in `pandas`.

Getting ready

We will work with the Amsterdam House Prices data for this recipe. You can retrieve all the files from the GitHub repository.

How to do it...

We will learn how to remove outliers and visualize the output using libraries such as `pandas`, `matplotlib`, and `seaborn`:

1. Import the relevant libraries:

    ```
    import pandas as pd
    import matplotlib.pyplot as plt
    import seaborn as sns
    ```

2. Load the `.csv` into a dataframe using `read_csv` and subset the dataframe to include the relevant columns:

    ```
    houseprice_data = pd.read_csv("data/HousingPricesData.csv")
    houseprice_data = houseprice_data[['Price', 'Area', 'Room']]
    ```

3. Check the first five rows using the `head` method. Check the number of columns and rows:

    ```
    houseprice_data.head()
          Price      Area    Room
    0     685000.0     64       3
    1     475000.0     60       3
    2     850000.0    109       4
    3     580000.0    128       6
    4     720000.0    138       5

    houseprice_data.shape
    (924, 3)
    ```

4. Identify missing values using the `isnull` method in pandas:

    ```
    houseprice_data.isnull().sum()
    Price    4
    Area     0
    Room     0
    ```

5. Drop missing values using the `dropna` method in pandas and save them in a new dataframe:

    ```
    houseprice_data_drop_missing = houseprice_data.dropna()
    houseprice_data_drop_missing.shape
    (920, 3)
    ```

6. Drop the missing values from a subset of columns using the `dropna` method in pandas:

    ```
    houseprice_data_drop_subset = houseprice_data.dropna(subset =
    ['Price', 'Area'])
    houseprice_data_drop_subset.shape
    (920, 3)
    ```

7.   Drop the entire column containing missing values using the `drop` method in `pandas`:

```
houseprice_data_drop_column = houseprice_data.
drop(['Area'],axis= 1)
houseprice_data_drop_column.shape
(924,2)
```

Great. We have just removed the missing values from our dataset.

How it works...

In *step 1*, we import the relevant libraries. In *step 2*, we use the `read_csv` method to load the house price data into a dataframe.

In *steps 3* and *4*, we inspect the dataset using the `head` method and `shape` attribute. We then use the `isnull` and `sum` methods to identify columns with missing values in the dataset. In *step 5*, we remove all rows that contain one or more missing values using the `dropna` method in `pandas`. This is listwise deletion. In *step 6*, we remove all rows that contain one or more missing values in the subset columns. We specify the columns in a list and supply this to the `subset` parameter of the `dropna` method in `pandas`. In *step 7*, we remove an entire column using the `drop` method in `pandas`. We specify the column name as the first argument and the axis as the second. An axis with a value of 1 represents dropping a column.

Replacing missing values

Removing missing values is a simple and quick approach to handling missing values. However, it is only effective when missing values are minimal. Replacing missing values is a better approach when there are many missing values within critical variables and when the values are missing at random. This approach is also called imputation.

We can fill in missing values using the following approaches:

- **Statistical measures**: This involves using summary statistics such as mean, median, percentiles, and so on.

- **Backfill or forward fill**: In sequential data, we can use the last value before the missing value or the next value after the missing value. This is known as backfill and forward fill, respectively. This method is more appropriate when dealing with time series data where the missing values are likely to be time dependent.

- **Model-based**: This involves using machine learning models such as linear regression or **K-nearest neighbors** (**KNN**). This is covered in the next recipe.

Generally, imputation methods are very useful. However, they can introduce bias when the missing data is not missing at random.

We will explore how to replace missing values using summary statistics. We will use the `mean`, `median` and `mode` methods in `pandas` to achieve this.

Getting ready

We will work with the Amsterdam House Prices data for this recipe. You can retrieve all the files from the GitHub repository.

How to do it...

We will learn how to replace missing values using the `pandas` library:

1. Import the relevant libraries:

    ```
    import pandas as pd
    ```

2. Load the `.csv` into a dataframe using `read_csv` and subset the dataframe to include the relevant columns:

    ```
    houseprice_data = pd.read_csv("data/HousingPricesData.csv")
    houseprice_data = houseprice_data[['Zip','Price', 'Area',
    'Room']]
    ```

3. Check the first five rows using the `head` method. Check the number of columns and rows:

    ```
    houseprice_data.head()
         Zip          Price       Area    Room
    0    1091 CR      685000.0     64      3
    1    1059 EL      475000.0     60      3
    2    1097 SM      850000.0     109     4
    3    1060 TH      580000.0     128     6
    4    1036 KN      720000.0     138     5

    houseprice_data.shape
    (924, 4)
    ```

4. Identify missing values using the `isnull` method in `pandas`:

    ```
    houseprice_data.isnull().sum()
    Zip      0
    Price    4
    Area     0
    Room     0
    ```

5. View the rows with missing values using the isnull method in pandas:

```
houseprice_data[houseprice_data['Price'].isnull()]
        Zip           Price      Area     Room
73      1017 VV                  147      3
321     1067 HP                  366      12
610     1019 HT                  107      3
727     1013 CK                  81       3
```

6. Compute replacement values using the mean, median, and mode methods in pandas:

```
mean = houseprice_data['Price'].mean()
median = houseprice_data['Price'].median()
mode = houseprice_data['Zip'].mode()[0]
print("mean: ",mean,"median: " ,median,"mode: ", mode)
mean:  622065 median:  467000 mode:  1075 XR
```

7. Replace the missing values with the mean:

```
houseprice_data['price_with_mean'] = houseprice_data['Price'].
fillna(mean)
houseprice_data.isnull().sum()
Zip                   0
Price                 4
Area                  0
Room                  0
price_with_mean       0
```

8. Replace the missing values with the median:

```
houseprice_data['price_with_median'] = houseprice_data['Price'].
fillna(median)
houseprice_data.isnull().sum()
Zip                   0
Price                 4
Area                  0
Room                  0
price_with_mean       0
price_with_median     0
```

9. Replace the missing values with the group mean and group median:

```
houseprice_data['group_mean'] = houseprice_data.groupby('Room')
['Price'].transform(lambda x: np.nanmean(x))
houseprice_data['group_median'] = houseprice_data.
groupby('Room')['Price'].transform(lambda x: np.nanmedian(x))
```

10. View all replacements:

```
houseprice_data[houseprice_data['Price'].isnull()]
```

This results in the following output:

	Zip	Price	Area	Room	price_with_mean	price_with_median	group_mean	group_median
73	1017 VV	NaN	147	3	622065.419565	467000.0	512416.39697	450000.0
321	1067 HP	NaN	366	12	622065.419565	467000.0	NaN	NaN
610	1019 HT	NaN	107	3	622065.419565	467000.0	512416.39697	450000.0
727	1013 CK	NaN	81	3	622065.419565	467000.0	512416.39697	450000.0

Figure 9.20: dataframe displaying the replaced missing values

Good. Now we have replaced the missing values using summary statistics.

How it works...

In *steps 1* and *2*, we import the pandas library and load the house price data into a dataframe using the read_csv method. We also select the relevant columns.

In *steps 3* to *5*, we inspect the dataset using the head method and shape attribute. We then identify columns with missing values and the number of missing values using the isnull and sum methods in pandas. We also get a glimpse of the missing values using the isnull method.

In *step 6*, we compute the replacement values using the mean, median, and mode methods in pandas. In *steps 7* and *8*, we use the fillna method to fill the missing values with the mean and median methods. Typically, the nature of the data and the objective of the analysis will determine whether the mean or median should be used to replace missing values. A key point to always remember is that the mean is sensitive to outliers while the median is not. For the mode, this is typically used for categorical variables. The fillna method can also be used to fill missing values in categorical variables using the value computed from the mode method.

In *step 9*, we create replacement values based on groupings. We use the groupby and transform methods in pandas and the nanmean and nanmedian methods in numpy to achieve this. The nanmean and nanmedian compute the mean and median by ignoring missing values (NaNs). The house with 12 rooms still gives a NaN because there is only one observation with 12 rooms and that observation has a missing price. Hence, there is no value to compute the average price of a 12-room house. In such a case, we may need to use a different method to replace this specific missing value. We can consider using machine learning models for this. This approach is covered in the next recipe.

This group replacement method typically yields more accurate results than using the average across all observations. However, it should be used when there is a relationship between other variables and the variable with the missing values. In step 10, we display all the replacements using the isnull method.

Imputing missing values using machine learning models

Beyond replacing missing values using statistical measures such as the mean, median, or percentiles, we can also use machine learning models to impute missing values. This process involves predicting the missing values based on the data available in other fields.

A very popular method is to use the K-Nearest Neighbor imputation (KNN). This involves identifying the k-nearest complete data points (neighbors) that surround the missing values and using the average of the values of these k-nearest data points to replace the missing values:

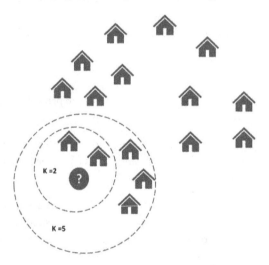

Figure 9.21: Illustration of KNN using house prices in a neighborhood

The preceding diagram gives a sense of how imputation works, specifically using the KNN algorithm. The price of the house with the question mark can be estimated based on the price of neighboring houses. In this example, we are using two immediate neighboring houses and five neighboring houses (**K =2** and **K =5**, respectively). There are several other models that can be used, such as linear regression, decision trees, and so on. However, we will focus on the KNN model in this recipe.

We will explore how to replace missing values using machine learning model values. We will use the KNNImputer class in scikit-learn to achieve this.

Getting ready

We will work with the Amsterdam House Prices data for this recipe. You can retrieve all the files from the GitHub repository.

How to do it...

We will learn how to replace missing values using the `pandas` and `scikit-learn` libraries:

1. Import the relevant libraries:

    ```
    import pandas as pd
    from sklearn.impute import KNNImputer
    ```

2. Load the `.csv` into a dataframe using `read_csv` and subset the dataframe to include the relevant columns:

    ```
    houseprice_data = pd.read_csv("data/HousingPricesData.csv")
    houseprice_data = houseprice_data [['Price', 'Area',
    'Room','Lon','Lat']]
    ```

3. Check the first five rows using the `head` method. Check the number of columns and rows:

    ```
    houseprice_data.head()
        Price       Area    Room    Lon         Lat
    0   685000.0    64      3       4.907736    52.356157
    1   475000.0    60      3       4.850476    52.348586
    2   850000.0    109     4       4.944774    52.343782
    3   580000.0    128     6       4.789928    52.343712
    4   720000.0    138     5       4.902503    52.410538

    houseprice_data.shape
    (924, 5)
    ```

4. Identify missing values using the `isnull` method in `pandas`:

    ```
    houseprice_data.isnull().sum()
    Price   4
    Area    0
    Room    0
    Lon     0
    Lat     0
    ```

5. Identify the index of missing values using the `index` method in `pandas` and save the index in a variable:

```
missing_values_index = houseprice_data[houseprice_data['Price'].
isnull()].index
missing_values_index
Int64Index([73, 321, 610, 727], dtype='int64')
```

6. Replace missing values with KNN imputed values:

```
imputer = KNNImputer(n_neighbors=5)
houseprice_data_knn_imputed = pd.DataFrame(imputer.fit_
transform(houseprice_data),columns = houseprice_data.columns)
```

7. View the output of the imputed dataset:

```
houseprice_data_knn_imputed.loc[missing_values_index,:]
```

This results in the following output:

	Price	Area	Room	Lon	Lat
73	1052000.0	147.0	3.0	4.897454	52.360707
321	3856000.0	366.0	12.0	4.787874	52.383877
610	694000.0	107.0	3.0	4.945022	52.369244
727	632000.0	81.0	3.0	4.880976	52.389623

Figure 9.22: dataframe displaying the imputed missing values

Well done. We just replaced missing values using a model-based approach.

How it works...

In *step 1*, we import the `pandas` and `scikit-learn` libraries. In *step 2*, we load the house price data into a dataframe using `read_csv`. We then subset for relevant columns.

In *step 3*, we inspect the dataset using the `head` method and `shape` attribute. In *step 4*, we check for columns with missing values using the `isnull` and `sum` methods. In *step 5*, we identify the index of the missing values using the `index` attribute in `pandas`.

In *step 6*, we create a model using the `KNNImputer` class in `scikit-learn` and replace the missing values using the `fit_transform` method in `scikit-learn`. In *step 8*, we display the output of the imputation.

10

Performing Automated Exploratory Data Analysis in Python

Sometimes, while analyzing large amounts of data, we may need to gain insights from the data very quickly. The insights typically gained could then form the basis for a detailed and in-depth analysis. Automated **Exploratory Data Analysis (EDA)** helps us to achieve this easily. Automated EDA automatically analyzes and visualizes data to extract trends, patterns, and insights with just a few lines of code. Automated EDA libraries typically carry out the cleaning, visualization, and statistical analysis of data in a quick and efficient way. These tasks would be difficult or time-consuming if they were performed manually.

Automated EDA is especially useful when dealing with complex or high-dimensional data, as finding relevant features can often be a daunting task. It is also helpful in reducing potential bias, especially in selecting the specific features to analyze in a large dataset.

In this chapter, we will cover Automated EDA using the various techniques detailed here:

- Doing Automated EDA using pandas profiling
- Performing Automated EDA using D-Tale
- Doing Automated EDA using AutoViz
- Performing Automated EDA using Sweetviz
- Implementing Automated EDA using custom code

Technical requirements

We will leverage the pandas, ydata_profiling, dtale, autoviz, sweetviz, matplotlib, and seaborn libraries in Python for this chapter. The code and notebooks for this chapter are available on GitHub at https://github.com/PacktPublishing/Exploratory-Data-Analysis-with-Python-Cookbook.

We will work within a virtual environment to avoid possible conflicts between some existing installations such as Jupyter Notebook and the installation of packages to be used in this chapter. In Python, a virtual environment is an isolated environment that has its own set of installed libraries, dependencies, and configurations. It is created to avoid conflicts between different versions of packages and libraries that are installed on a system. When we create a virtual environment, we can install specific versions of packages and libraries without affecting the global installation on our system.

The following steps and commands can be used to set this up:

1. Open the command prompt within your working directory and run the following command to create a Python environment called myenv:

    ```
    python -m venv myenv
    ```

2. Initialize the environment using the following command:

    ```
    myenv\Scripts\activate.bat
    ```

3. Install all the relevant libraries for the chapter:

    ```
    pip install pandas jupyter pandas_profiling dtale autoviz
    sweetviz seaborn
    ```

Doing Automated EDA using pandas profiling

pandas profiling is a popular Automated EDA library that generates EDA reports from a dataset stored in a pandas dataframe. With a line of code, the library can generate a detailed report, which covers critical information such as summary statistics, distribution of variables, correlation/interaction between variables, and missing values. The library is useful for quickly and easily generating insights from large datasets because it requires minimal effort from its users. The output is presented in an interactive HTML report, which can easily be customized.

The Automated EDA report generated by pandas profiling contains the following sections:

*   **Overview**: This section provides a general summary of the dataset. It includes the number of observations, the number of variables, missing values, duplicate rows, and more.

- **Variables**: This section provides information about the variables in the dataset. It includes summary statistics (mean, median, standard deviation, and distinct values), data type, missing values, and more. It also has a subsection that provides histograms, common values, and extreme values.

- **Correlations**: This section provides a correlation matrix using a heatmap to highlight the relationship between all the variables in the dataset.

- **Interactions**: This section shows the bivariate analysis between variable pairs in the dataset. It uses scatterplots, density plots, and other visualizations to perform this. Users can switch between variables at the click of a button.

- **Missing values**: This section highlights the number and percentage of missing values that exist in each variable. It also provides information using a matrix, heatmap, and dendrogram.

- **Sample**: This section provides a sample of the dataset containing the first 10 and last 10 rows.

The following are some customizations we can achieve in pandas profiling:

- We can use the minimal configuration parameter to exclude expensive computations such as correlations or interactions between variables. This is very useful when working with large datasets.

- We can provide additional metadata or information about our dataset and its variables so that this is also displayed under the overview section. This is critical if we plan to share the report with the public or a team.

- We can store the profile report as an HTML file using the `to_file` method.

In this recipe, we will explore using the `ProfileReport` class in the `ydata_profiling` library to generate an Automated EDA report. The `Pandas_profiling` library has been renamed to `ydata_profiling`. However, it is still popularly known as the former.

Getting ready

We will work with only one dataset in this chapter: **Customer Personality Analysis data** from Kaggle.

Create a folder specifically for this chapter. Within this folder, create a new Python script or Jupyter Notebook file and create a subfolder called `data`. In the subfolder, place the `marketing_campaign.csv` file.

Alternatively, you could obtain all the necessary files from the GitHub repository.

Note

Kaggle provides the Customer Personality Analysis data for public use at `https://www.kaggle.com/datasets/imakash3011/customer-personality-analysis`.

In this chapter, we will use the full dataset for the different recipes. The data is also available in the repository for this book.

How to do it...

We will learn how to perform automated EDA using the pandas and ydata_profiling libraries:

1. Import the pandas and ydata_profiling libraries:

    ```
    import pandas as pd
    from ydata_profiling import ProfileReport
    ```

2. Load the .csv into a dataframe using read_csv:

    ```
    marketing_data = pd.read_csv("data/marketing_campaign.csv")
    ```

3. Create an Automated EDA report using the ProfileReport class in the ydata_profiling library. Use the to_file method to output the report to an HTML file:

    ```
    profile = ProfileReport(marketing_data)
    profile.to_file("Reports/profile_output.html")
    ```

    This results in the following output:

    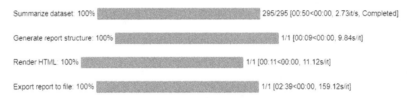

    Figure 10.1: The Pandas Profiling Report progress bar

4. Open the HTML output file in the Reports directory and view the report's **Overview** section:

Figure 10.2: Pandas Profiling Report Overview

5. View the report's **Variables** section:

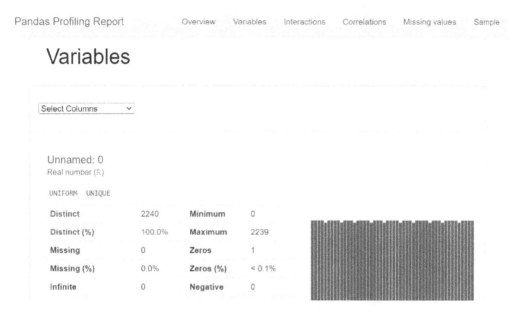

Figure 10.3: Pandas Profiling Report Variables

6. View the variable **Interactions** section:

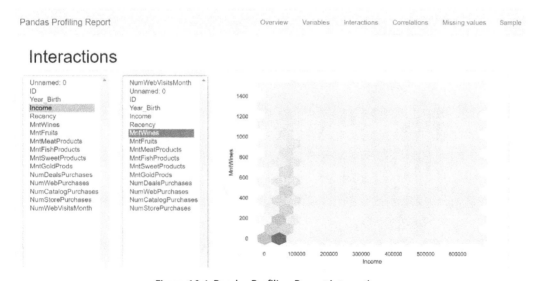

Figure 10.4: Pandas Profiling Report Interactions

7. View the **Correlations** section:

Figure 10.5: Pandas Profiling Report Correlations

8. View the **Missing values** section:

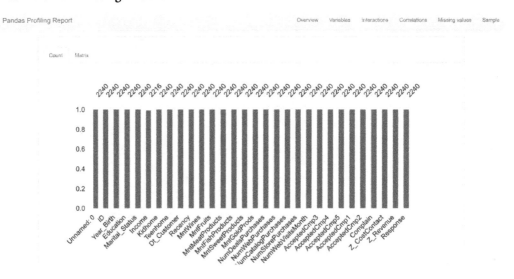

Figure 10.6: Pandas Profiling Report Missing values

9.   Create an Automated EDA report with the minimal configuration parameter to exclude expensive computations such as correlations or interactions between variables:

```
profile_min = ProfileReport(marketing_data, minimal=True)
profile_min.to_file("Reports/profile_minimal_output.html")
```

This results in the following output:

Figure 10.7: The Pandas Profiling Report progress bar

10.   Open the HTML output file in the Reports directory and view the report's **Overview** section:

Figure 10.8: Pandas Profiling Report minimal report

11. Create an Automated EDA report with additional metadata or information about our dataset and its variables:

```
profile_meta = ProfileReport(
    marketing_data,
    title="Customer Personality Analysis Data",
    dataset={
        "description": "This data contains marketing and sales
data of a company's customers. It is useful for identifying the
most ideal customers to target.",
        "url": "https://www.kaggle.com/datasets/imakash3011/
customer-personality-analysis.",
    },
    variables= {
        "descriptions": {
            "ID": "Customer's unique identifier",
            "Year_Birth": "Customer's birth year",
            "Education": "Customer's education level",
            ...         ...          ...         ...         ...
            "MntSweetProducts": "Amount spent on sweets in last
2 years",
            "MntGoldProds": "Amount spent on gold in last 2
years",
        }
    }
)

profile_meta.to_file("Reports/profile_with_metadata.html")
```

This results in the following output:

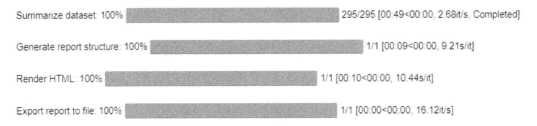

Figure 10.9: The Pandas Profiling Report progress bar

12. Open the HTML output file in the Reports directory and view the report's **Overview** and **Variables** pages.

The **Overview** section displays the following information:

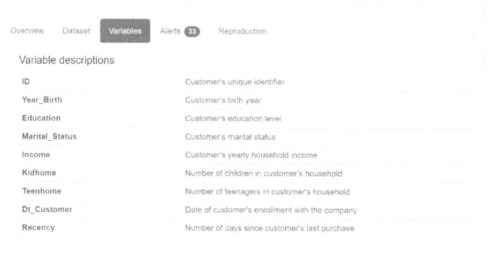

Figure 10.10: Pandas Profiling Report Dataset tab with metadata

The **Variables** section displays the following detail:

Customer Personality Analysis Data   Overview   Variables   Interactions   Correlations   Missing values   Sample

Overview

Overview   Dataset   Variables   Alerts 33   Reproduction

Variable descriptions

ID	Customer's unique identifier
Year_Birth	Customer's birth year
Education	Customer's education level
Marital_Status	Customer's marital status
Income	Customer's yearly household income
Kidhome	Number of children in customer's household
Teenhome	Number of teenagers in customer's household
Dt_Customer	Date of customer's enrollment with the company
Recency	Number of days since customer's last purchase

Figure 10.11: Pandas Profiling Report Variables tab with metadata

Great. We just performed Automated EDA on our dataset using pandas profiling.

How it works...

In this recipe, we use the `pandas` and `ydata_profiling` libraries. In step 1, we import these libraries. In step 2, we load the Customer Personality Analysis data using the `read_csv` method.

In step 3, we use the `ProfileReport` class in `ydata_profiling` to generate an Automated EDA report on our data. We also use the `to_file` method to export the report to an HTML file in the `Reports` folder in our active directory. In step 4, we open `profile_output.html` in the `Reports` folder and view the report's **Overview** section. The section contains tabs that cover summary statistics about our data and alert on some issues with the data, such as missing values, zero values, unique values, and more. Lastly, it contains a **Reproduction** tab that covers information about the report generation. At the top right of the page, we can see tabs for all the sections available in the report (**Overview, Variables, Interactions, Correlations**, and more); we can navigate to each section by clicking on the relevant tab or scrolling down the report.

In step 5, we go to the **Variables** section, which contains summary statistics of each variable. In step 6, we view the variable **Interactions** section, which shows the relationships between our numerical variables. We can select variables on the left and right pane to view various interactions. In step 7, we view the **Correlations** section, which quantifies the relationship between variables. This section provides a **Heatmap** tab and a **Table** tab to display the correlation.

In step 8, we view the **Missing values** section. This shows the number of missing values within all variables. It provides this information in either a bar chart or a matrix chart.

In steps 9 and 10, we use the `ProfileReport` class again to generate an Automated EDA report with minimal information. For the `minimal` parameter, we specify the value as `True` and then export the report to an HTML file using the `to_file` method. We then open `profile_minimal_output.html` in the `Reports` folder and view the report's **Overview** section. At the top right of the page, we can see only two tabs available in the report (**Overview** and **Variables**). The minimal configuration is useful when working with very large datasets, and there is a need to exclude expensive computations such as correlations or interactions between variables.

In step 11, we use the `ProfileReport` class again to generate an Automated EDA report with the metadata of our dataset. We provide the dataset title in the `title` parameter; we provide the description of the dataset as a dictionary in the `dataset` parameter, and we provide the variable descriptions as another dictionary in the `variables` parameter. In step 12, we view the report. The report now has a **Dataset and Variables** subtab under the **Overview** section. These tabs display the metadata of our dataset.

See also...

You can check out these useful resources on pandas profiling:

- `https://pub.towardsai.net/advanced-eda-made-simple-using-pandas-profiling-35f83027061a`

- https://ydata-profiling.ydata.ai/docs/master/pages/advanced_
  usage/available_settings.html

Performing Automated EDA using dtale

The dtale library is an Automated EDA library that provides a graphical interface for analyzing pandas dataframes. Users can easily interact with the data and perform common tasks such as filtering, sorting, grouping, and visualizing data to generate quick insights. It speeds up the EDA process on large datasets. Beyond EDA, it provides support for common data cleaning tasks such as handling missing values and removing duplicates. It has a web-based interface that makes interaction with data more intuitive and less tedious. It provides a shareable link for the analysis performed.

While pandas profiling does the heavy lifting by providing a detailed report, Dtale, on the other hand, provides a flexible user interface for users to perform several cleaning and EDA tasks. With Dtale, users can edit cells, perform conditional formatting, perform operations on columns, generate summary statistics, and create various visualizations. Dtale typically provides flexibility but requires some effort from its user.

In this recipe, we will explore common options available in the dtale library for Automated EDA.

Getting ready

We will continue working with the **Customer Personality Analysis data** from Kaggle in this recipe. You can retrieve all the files from the GitHub repository.

How to do it...

We will learn how to perform Automated EDA using libraries such as pandas and dtale with the help of the following steps:

1. Import the relevant libraries:

    ```
    import pandas as pd
    import dtale
    ```

2. Load the .csv into a dataframe using read_csv:

    ```
    marketing_data = pd.read_csv("data/marketing_campaign.csv")
    ```

3. Use the show and open_browser methods in dtale to display the data in the browser. The show method alone will display the data in Jupyter Notebook only:

    ```
    dtale.show(marketing_data).open_browser()
    ```

This results in the following output:

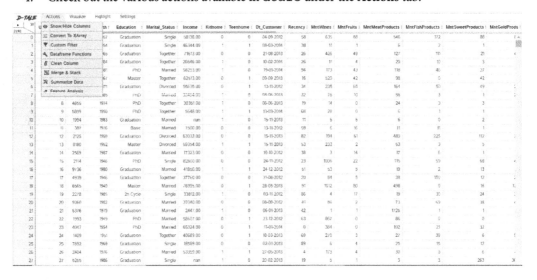

Figure 10.12: The Dtale Data page

4. Check out the various actions available in dtale under the **Actions** tab:

Figure 10.13: The Dtale Actions tab

5. Check out the various visualization options available in dtale under the **Visualize** tab. Also, check out the options under the **Highlight** tab.

The **Visualize** tab displays the following options:

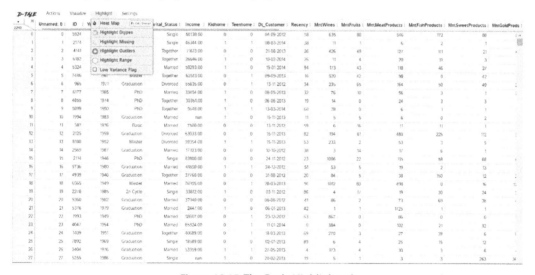

Figure 10.14: The Dtale Visualize tab

The **Highlight** tab displays the following options:

Figure 10.15: The Dtale Highlight tab

6.   Summarize the dataset using the **Summarize Data** option under the **Actions** tab.

The **Summarize Data** option displays the following parameters:

Figure 10.16: The Summarize Data parameters window

The **Summarize Data** parameters display the following results:

	Marital_Status	2n Cycle	Basic	Graduation	Master	PhD
0	Absurd	nan	nan	79244.00	65487.00	nan
1	Alone	nan	nan	34176.00	61331.00	35860.00
2	Divorced	1136088.00	9548.00	6488599.00	1862282.00	2761024.00
3	Married	3696088.00	439210.00	21793311.00	7353472.00	11046226.00
4	Single	1932262.00	328296.00	12625257.00	4014792.00	5118203.00
5	Together	2505239.00	297361.00	15891167.00	5315119.00	6500805.00
6	Widow	256961.00	22123.00	1924183.00	642417.00	1446914.00
7	YOLO	nan	nan	nan	nan	96864.00

Figure 10.17: The Summarize Data dataset

7. Describe the dataset using the **Describe** option under the **Visualize** tab:

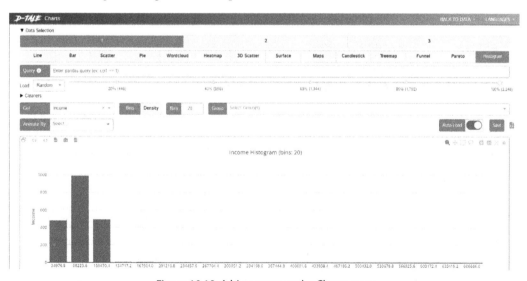

Figure 10.18: The Describe page

8. Plot a histogram using the **Charts** option under the **Visualize** tab:

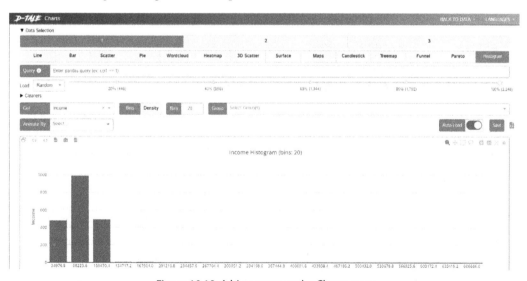

Figure 10.19: A histogram on the Charts page

Awesome. We just performed Automated EDA using the `dtale` library.

How it works...

In step 1, we import the relevant `pandas` and `dtale` libraries. In step 2, we load the Customer Personality Analysis data into a `pandas` dataframe using `read_csv`.

In steps 3 to 5, we use the `show` method in `dtale` to display our data and the various navigation options the library contains for data cleaning and EDA. We also view the options available under the **Actions**, **Visualize**, and **Highlight** tabs. In step 6, we summarize our data using the **Summarize** option under the **Actions** tab. This creates a modal containing the parameters for the **Summarize** option. We select the **Pivot** operation and specify the row, columns, values, and aggregation operation using the **Operation**, **Rows**, **Columns**, **Values**, and **Aggregation** parameters, respectively. The output is provided in a new tab in the browser.

In step 7, we generate summary statistics of our data using the **Describe** option under the **Visualize** tab. This creates a new tab containing all the columns on the left and summary statistics on the right. In step 8, we create a histogram using the **Charts** option under the **Visualize** tab. This creates a new tab with the various charts. We select the data instance under the **Data Selection** parameter; we select the **Histogram** chart option under the list of charts and then provide the column name and the number of bins under the **Col** and **Bins** parameters. We also set the **Load** parameter to **100%** to ensure that 100% of the data is displayed. The output chart provides various options to export and interact with the chart.

See also

You can check out these useful resources on Dtale:

- `https://dtale.readthedocs.io/en/latest/`

Doing Automated EDA using AutoViz

AutoViz is an Automated EDA library used for the automatic visualization of datasets. Unlike the previous libraries, it is built on top of the `matplotlib` library. It provides a wide array of visuals to summarize and analyze datasets to provide quick insights. The library does most of the heavy lifting and requires minimal user input.

The reports generated by the AutoViz library typically provide the following:

- **Data cleaning suggestions**: They provide insights into missing values, unique values, and outliers. They also provide suggestions on how to handle outliers, irrelevant columns, rare categories, columns with constant values, and more. This can be useful for data cleaning.

- **Univariate analysis**: They use histograms, density plots, and violin plots to provide insights into the distribution of the data, outliers, and more.

- **Bivariate analysis**: They use scatterplots, heatmaps, and pair plots to provide insights into the relationship between two variables in the dataset.

- **Correlation analysis**: They compute the correlation between variables in the dataset and visualize them using heatmaps to display correlation strength and direction.

We will explore how to perform Automated EDA using the `AutoViz_Class` class in the `autoviz` library.

Getting ready

We will work with the Customer Personality Analysis data from Kaggle on this recipe. You can retrieve all the files from the GitHub repository.

How to do it...

We will learn how to identify bivariate outliers using scatterplots and boxplots. We will use libraries such as `pandas` and `autoviz`:

1. Import the relevant libraries:

```
import pandas as pd
from autoviz.AutoViz_Class import AutoViz_Class
Imported v0.1.601. After importing, execute '%matplotlib inline'
to display charts in Jupyter.
    AV = AutoViz_Class()
    dfte = AV.AutoViz(filename, sep=',', depVar='', dfte=None,
header=0, verbose=1, lowess=False,
                chart_format='svg',max_rows_analyzed=150000,max_
cols_analyzed=30, save_plot_dir=None)
Update: verbose=0 displays charts in your local Jupyter
notebook.
        verbose=1 additionally provides EDA data cleaning
suggestions. It also displays charts.
        verbose=2 does not display charts but saves them in
AutoViz_Plots folder in local machine.
        chart_format='bokeh' displays charts in your local
Jupyter notebook.
        chart_format='server' displays charts in your browser:
one tab for each chart type
        chart_format='html' silently saves interactive HTML
files in your local machine
```

2. Load the .csv into a dataframe using read_csv:

```
marketing_data = pd.read_csv("data/marketing_campaign.csv")
```

3. Execute the following code to display charts inside Jupyter Notebook. The inline command ensures the charts are displayed in Jupyter Notebook:

```
%matplotlib inline
```

4. Create an Automated EDA report using the AutoViz_Class class in autoviz:

```
viz = AutoViz_Class()
df = viz.AutoViz(filename = '',dfte= marketing_data, verbose =1)
```

This results in the following output:

```
Shape of your Data Set loaded: (2240, 30)
#################################################################################
####################### C L A S S I F Y I N G  V A R I A B L E S ###################
#################################################################################
Classifying variables in data set...
Data cleaning improvement suggestions. Complete them before proceeding to ML modeling.
```

	Nullpercent	NuniquePercent	dtype	Nuniques	Nulls	Least num. of categories	Data cleaning improvement suggestions
Income	1.071429	88.125000	float64	1974	24	0	fill missing values, highly right skewed distribution: drop outliers or do box-cox transform
Unnamed: 0	0.000000	100.000000	int64	2240	0	0	possible ID column: drop
NumDealsPurchases	0.000000	0.669643	int64	15	0	0	
Z_Revenue	0.000000	0.044643	int64	1	0	0	invariant values: drop
Z_CostContact	0.000000	0.044643	int64	1	0	0	invariant values: drop
Complain	0.000000	0.089286	int64	2	0	0	
AcceptedCmp2	0.000000	0.089286	int64	2	0	0	
AcceptedCmp1	0.000000	0.089286	int64	2	0	0	
AcceptedCmp5	0.000000	0.089286	int64	2	0	0	
AcceptedCmp4	0.000000	0.089286	int64	2	0	0	
AcceptedCmp3	0.000000	0.089286	int64	2	0	0	
NumWebVisitsMonth	0.000000	0.714286	int64	16	0	0	
NumStorePurchases	0.000000	0.625000	int64	14	0	0	
NumCatalogPurchases	0.000000	0.625000	int64	14	0	0	
NumWebPurchases	0.000000	0.669643	int64	15	0	0	
MntGoldProds	0.000000	9.508929	int64	213	0	0	
ID	0.000000	100.000000	int64	2240	0	0	possible ID column: drop
MntSweetProducts	0.000000	7.901786	int64	177	0	0	
MntFishProducts	0.000000	8.125000	int64	182	0	0	
MntMeatProducts	0.000000	24.910714	int64	558	0	0	
MntFruits	0.000000	7.053571	int64	158	0	0	
MntWines	0.000000	34.642857	int64	776	0	0	
Recency	0.000000	4.464286	int64	100	0	0	

Figure 10.20: Data cleaning suggestions

The following is a sample of some of the univariate charts generated:

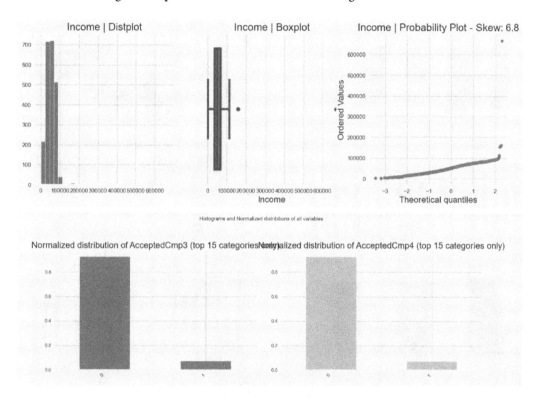

Figure 10.21: Univariate analysis (a sample of charts)

The following is a sample of some of the bivariate charts generated:

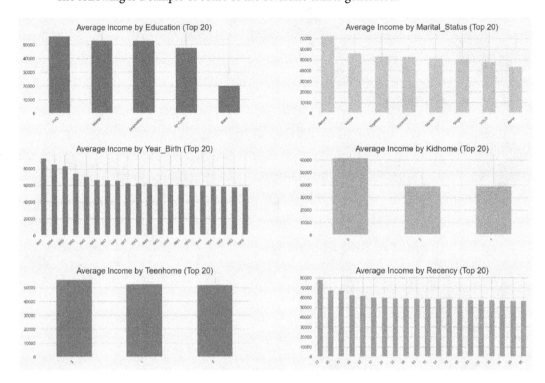

Figure 10.22: Bivariate analysis (a sample of charts)

The following is the correlation heatmap that is generated:

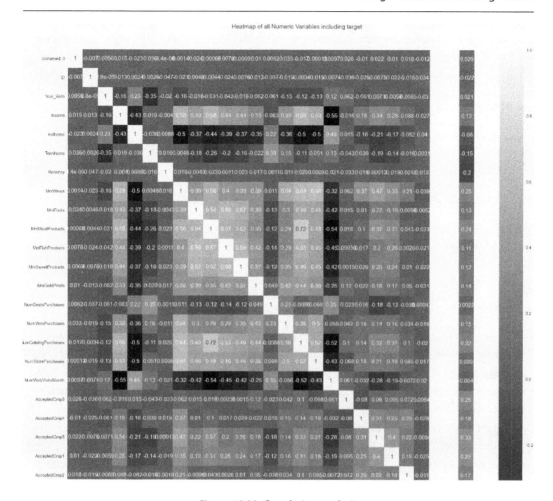

Figure 10.23: Correlation analysis

Well done. We just performed Automated EDA using `autoviz`.

How it works...

In steps 1 and 2, we import the `pandas` and `autoviz` libraries and load our Customer Personality Analysis data from a `.csv` file into a dataframe using the `read_csv` method.

In step 3, we execute the `matplotlib inline` command to ensure that charts are displayed inside Jupyter Notebook. In step 4, we use `AutoViz_Class` to generate an Automated EDA report. We provide an empty string to the `filename` parameter since we aren't loading a `.csv` file directly. We provide our dataframe variable to the `dfte` parameter since a dataframe is our data source. Lastly, we provide a value of `1` to the `verbose` parameter to ensure the report contains all charts and more information.

The output of the code provides data cleaning suggestions on handling missing values, skewed data, outliers, and irrelevant columns. It also contains bar charts, boxplots, and histograms for univariate analysis. Lastly, it contains correlation analysis, scatterplots, bar charts, and more for bivariate analysis.

See also

You can check out these useful resources:

- `https://github.com/AutoViML/AutoViz`

Performing Automated EDA using Sweetviz

Sweetviz is another Automated EDA library that generates visualizations on datasets automatically. It can provide an in-depth analysis of a dataset and can also be used to compare datasets. Sweetviz provides statistical summaries and visualizations in an HTML report like the `pandas` profiling library. Like its counterparts, it speeds up the EDA process and requires minimal user input.

The Sweetviz report typically contains the following elements:

- **Overview**: This provides a summary of the dataset. This includes the number of observations, missing values, summary statistics (mean, median, max, and min), data types, and more.
- **Associations**: This generates a heatmap that visualizes the correlation between variables, both categorical and numerical. The heatmap provides insights into the strength and direction of the association.
- **Target analysis**: This provides insights into how a target variable is influenced by other variables once we specify the target variable as an argument in the `sweetviz` report methods.
- **Comparative analysis**: This provides the functionality to compare two datasets side by side. It generates visualizations and statistical summaries to identify the differences between them. This can be very useful for machine learning when we need to compare the train data and test data.

We will now explore how to perform Automated EDA using the `analyze` and `compare` methods in `sweetviz`.

Getting ready

We will work with the Customer Personality Analysis data from Kaggle on this recipe. You can retrieve all the files from the GitHub repository.

How to do it...

We will learn how to spot multivariate outliers using libraries such as `pandas` and `sweetviz`:

1. Import the `pandas` and `sweetviz` libraries:

```
import pandas as pd
import sweetviz
```

2. Load the .csv into a dataframe using read_csv:

```
marketing_data = pd.read_csv("data/marketing_campaign.csv")
```

3. Create an Automated EDA report using the analyze method in the sweetviz library. Use the show_html method to export the report to an HTML file:

```
viz_report = sweetviz.analyze(marketing_data, target_feat = 'Response')
viz_report.show_html('Reports/sweetviz_report.html')
```

This results in the following output:

Figure 10.24: The Sweetviz progress bar

4. View the report generated in a new browser tab:

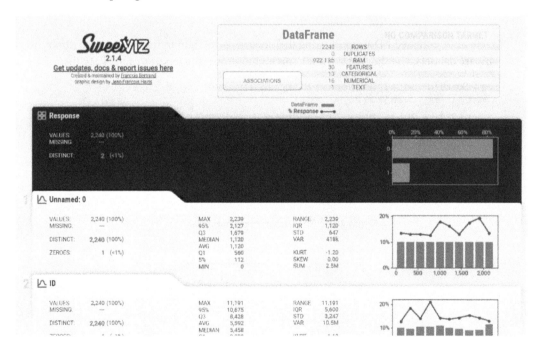

Figure 10.25: Sweetviz overview and target analysis

The **Associations** button generates the following:

Figure 10.26: Sweetviz Associations

5. Create an Automated EDA report using the `compare` method in the `sweetviz` library to compare datasets. Use the `show_html` method to export the report to an HTML file:

```
marketing_data1 = sweetviz.compare(marketing_data[1220:],
marketing_data[:1120])
marketing_data1.show_html('Reports/sweetviz_compare.html')
```

This results in the following output:

Done! Use 'show' commands to display/save.   [100%] 00:04 -> (00:00 left)

Report Reports/sweetviz_compare.html was generated! NOTEBOOK/COLAB USERS: the web browser MAY not pop up, regardless, the repor
t IS saved in your notebook/colab files.

Figure 10.27: The Sweetviz progress bar

6. View the report generated in a new browser tab:

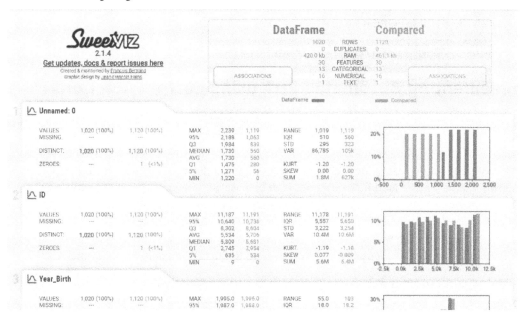

Figure 10.28: Sweetviz associations

Good job. Now we have performed Automated EDA using `sweetviz`.

How it works...

In steps 1 and 2, we import the `pandas` and `sweetviz` libraries. We also load our Customer Personality Analysis data from a `.csv` file into a dataframe.

In steps 3 and 4, we create an Automated EDA report using the `analyze` method in the `sweetviz` library. We use the `target_feat` parameter to specify the column to be used as a target for target analysis. We also provide the report output in HTML using the `show_html` method. We then view the report in a new browser tab.

In the report, we first see summary statistics of the dataframe. We then see summary statistics for each variable and an analysis of each variable against the target variable. By selecting the **Associations** button at the top, the associations plot is generated to the right showing how categorical and numerical values are related. The squares represent categorical associations (uncertainty coefficient and correlation ratio) between 0 and 1, while the circles represent numerical correlations (Pearson's) between 0 and 1.

In steps 5 and 6, we create another automated report to compare subsets of our dataset. We use the `compare` method to compare subsets of our dataset. The output of the report shows a comparison between the two datasets across all the variables.

See also

You can check out these useful resources:

- `https://pypi.org/project/sweetviz/`

Implementing Automated EDA using custom functions

Sometimes, when performing Automated EDA, we may require more flexibility around the visual options and analysis techniques. In such cases, custom functions may be preferable to the libraries discussed in the preceding recipes since they may be limited in providing such flexibility. We can write custom functions using the preferred visual options and analysis techniques and save them as a Python module. The module ensures our code is reusable, which means we can easily perform EDA automatically without having to write several lines of code. This gives us significant flexibility, especially because the module can be constantly improved based on preferences. Most of the heavy lifting happens only when we are initially writing the custom functions. Once they have been saved into a module, they become reusable, and we only need to call the functions in single lines of code.

We will explore how to write custom functions to perform Automated EDA. We will use the `matplotlib`, `seaborn`, `Ipython`, `itertools`, and `pandas` libraries to achieve this.

Getting ready

We will work with the Customer Personality Analysis data from Kaggle in this recipe. You can retrieve all the files from the GitHub repository.

How to do it...

We will learn how to handle outliers using the flooring and capping approach. We will use libraries such as `matplotlib`, `seaborn`, `Ipython`, `itertools` and `panda`:

1. Import the relevant libraries:

   ```
   import pandas as pd
   import matplotlib.pyplot as plt
   import seaborn as sns
   from IPython.display import HTML
   from itertools import combinations, product
   ```

2. Load the `.csv` into a dataframe using `read_csv`:

   ```
   marketing_data = pd.read_csv("data/marketing_campaign.csv")
   ```

3. Create a custom summary statistics function that displays summary stats for categorical and numerical variables side by side:

```
def dataframe_side_by_side(*dataframes):
    html = '<div style="display:flex">'
    for dataframe in dataframes:
        html += '<div style="margin-right: 2em">'
        html += dataframe.to_html()
        html += '</div>'
    html += '</div>'
    display(HTML(html))

def summary_stats_analyzer(data):
    df1 = data.describe(include='object')
    df2 = data.describe()
    return dataframe_side_by_side(df1,df2)
```

4. Generate summary statistics using the custom function:

```
summary_stats_analyzer(marketing_data)
```

This results in the following output:

	Education	Marital_Status	Dt_Customer		Unnamed: 0	ID	Year_Birth	Income	Kidhome	Teenhome	Recency	
count	2240	2240	2240	count	2240.000000	2240.000000	2240.000000	2216.000000	2240.000000	2240.000000	2240.000000	22
unique	5	8	663	mean	1119.500000	5592.159821	1968.805804	52247.251354	0.444196	0.506250	49.109375	3
top	Graduation	Married	31-08-2012	std	646.776623	3246.662198	11.984069	25173.076661	0.538398	0.544538	28.962453	3
freq	1127	864	12	min	0.000000	0.000000	1893.000000	1730.000000	0.000000	0.000000	0.000000	
				25%	559.750000	2828.250000	1959.000000	35303.000000	0.000000	0.000000	24.000000	
				50%	1119.500000	5458.500000	1970.000000	51381.500000	0.000000	0.000000	49.000000	1
				75%	1679.250000	8427.750000	1977.000000	68522.000000	1.000000	1.000000	74.000000	5
				max	2239.000000	11191.000000	1996.000000	666666.000000	2.000000	2.000000	99.000000	14

Figure 10.29: Summary statistics

5. Identify numerical and categorical data types within the data. This is a requirement for univariate analysis, given various chart options:

```
categorical_cols = marketing_data.select_dtypes(include =
'object').columns
categorical_cols[:4]
```
**Index(['Education', 'Marital_Status', 'Dt_Customer'],
dtype='object')**

```
discrete_cols = [col for col in marketing_data.select_
dtypes(include = 'number') if marketing_data[col].nunique() <
15]
```

```
discrete_cols[:4]
['Kidhome', 'Teenhome', 'NumCatalogPurchases',
'NumStorePurchases']


numerical_cols = [col for col in marketing_data.select_
dtypes(include = 'number').columns if col not in discrete_cols]
numerical_cols[:4]
['Unnamed: 0', 'ID', 'Year_Birth', 'Income']
```

6. Create a custom function for performing univariate analysis:

```
def univariate_analyzer (data,subset):
categorical_cols = data.select_dtypes(include = 'object').
columns
    discrete_cols = [col for col in marketing_data.select_
dtypes(include = 'number') if marketing_data[col].nunique() <
15]
    numerical_cols = [col for col in marketing_data.select_
dtypes(include = 'number').columns if col not in discrete_cols]
    all_cols = data.columns

    plots = []
    if subset == 'cat':
        print("categorical variables: ", categorical_cols)
        for i in categorical_cols:
            plt.figure()
            chart = sns.countplot(data = data, x= data[i])
            plots.append(chart)
...       ...        ...           ...            ...
    else:
        for i in all_cols:
            if i in categorical_cols:
                plt.figure()
                chart = sns.countplot(data = data, x= data[i])
                plots.append(chart)
...       ...        ...           ...          ...

            else:
                pass

    for plot in plots:
        print(plot)
```

7. Perform univariate analysis using the custom code:

```
univariate_analyzer(marketing_data,'all')
```

This results in several charts; the following output is a subset of these:

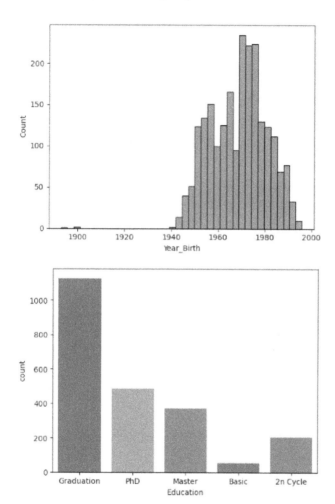

Figure 10.30: Custom EDA function univariate analysis (a sample of charts)

8. Create a subset of the original data to avoid creating too many bivariate charts:

```
marketing_sample = marketing_data[['Education', 'Marital_
Status', 'Income', 'Kidhome', 'MntWines',
                                   'MntMeatProducts',
 'NumWebPurchases', 'NumWebVisitsMonth','Response']]
```

9. Identify numerical and categorical data types within the subset data. This is a requirement for bivariate analysis, given various chart options:

```
categorical_cols_ = marketing_sample.select_dtypes(include =
'object').columns
discrete_cols_ = [col for col in marketing_sample.select_
dtypes(include = 'number') if marketing_sample[col].nunique() <
15]
numerical_cols_ = [col for col in marketing_sample.select_
dtypes(include = 'number').columns if col not in discrete_cols_]
```

10. Create value pairs for all column data types such as numerical-categorical, numerical-numerical, and more:

```
num_cat = [(i,j) for i in numerical_cols_ for j in categorical_
cols_ ]
num_cat[:4]
[('Income', 'Education'),
 ('Income', 'Marital_Status'),
 ('MntWines', 'Education'),
 ('MntWines', 'Marital_Status')]


cat_cat = [t for t in combinations(categorical_cols_, 2)]
cat_cat[:4]
[('Education', 'Marital_Status')]


num_num = [t for t in combinations(numerical_cols_, 2)]
num_num[:4]
[('Income', 'MntWines'),
 ('Income', 'MntMeatProducts'),
 ('Income', 'NumWebPurchases'),
 ('Income', 'NumWebVisitsMonth')]


dis_num = [(i,j) for i in discrete_cols_ for j in numerical_
cols_ if i != j]
dis_num[:4]
[('Kidhome', 'Income'),
 ('Kidhome', 'MntWines'),
 ('Kidhome', 'MntMeatProducts'),
 ('Kidhome', 'NumWebPurchases')]


dis_cat = [(i,j) for i in discrete_cols_ for j in categorical_
cols_ if i != j]
dis_cat[:4]
[('Kidhome', 'Education'),
```

```
  ('Kidhome', 'Marital_Status'),
  ('Response', 'Education'),
  ('Response', 'Marital_Status')]
```

11. Create a subset of the original data to avoid creating too many bivariate charts:

```
def bivariate_analyzer (data):
    categorical_cols_ = marketing_sample.select_dtypes(include =
'object').columns
    discrete_cols_ = [col for col in marketing_sample.select_
dtypes(include = 'number') if marketing_sample[col].nunique() <
15]
    numerical_cols_ = [col for col in marketing_sample.select_
dtypes(include = 'number').columns if col not in discrete_cols]

    num_num = [t for t in combinations(numerical_cols_, 2)]
    cat_cat = [t for t in combinations(categorical_cols_, 2)]
    num_cat = [(i,j) for i in numerical_cols_ for j in
categorical_cols_ ]
    dis_num = [(i,j) for i in discrete_cols_ for j in numerical_
cols_ if i != j]
    dis_cat = [(i,j) for i in discrete_cols_ for j in
categorical_cols_ if i != j]


    plots = []
    for i in num_num:
        plt.figure()
        chart = sns.scatterplot(data = data, x= data[i[0]], y=
data[i[1]])
        plots.append(chart)
    for i in num_cat:
        plt.figure()
        chart = sns.boxplot(data = data, x= data[i[1]], y=
data[i[0]] )
        plots.append(chart)
...          ...          ...          ...          ...

    else:
        pass

    for plot in plots:
        print(plot)
```

12. Perform bivariate analysis on the subset of data:

```
bivariate_analyzer(marketing_sample)
```

This results in several charts; the following output is a subset of these:

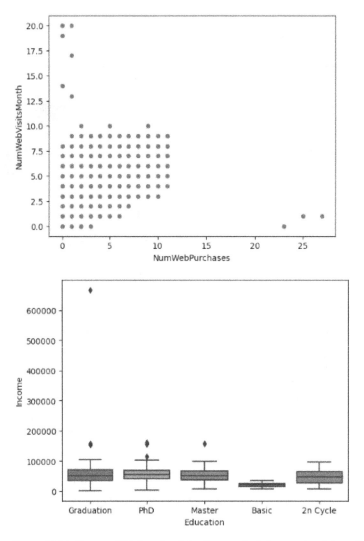

Figure 10.31: Custom EDA function bivariate analysis (a sample of charts)

Good. Now we have performed Automated EDA using custom code.

How it works...

In step 1, we import the `matplotlib`, `seaborn`, `Ipython`, `itertools`, and `pandas` libraries. In step 2, we load the Customer Personality Analysis data into a dataframe. In steps 3 and 4, we create a `dataframe_side_by_side` custom function that generates two dataframes side by side using the `to_html` method in `pandas` and the HTML class in `Ipython`. In the function, we define the style and use the addition assignment to add more styles and dataframes. Finally, we provide the HTML output. We then create another `summary_stats_analyzer` custom function that outputs the summary statistics for the object and numerical datatypes. We use the `describe` method in `pandas` to achieve this. For the summary stats of the object variables, we use the `include` parameter to specify this, while for the numerical variables, we use the default parameter. We generate summary stats by running our custom function on the data.

In steps 5 to 7, we identify numerical and categorical data types within the data. This is a requirement for univariate analysis, given various chart options that depend on data types. We use the `select_dtypes` method in `unique` to identify numerical or object data types. Because discrete numerical values are better displayed using bar charts, we identify discrete columns with less than 15 unique values using the `unique` method in `pandas`. We also exclude discrete columns from our numerical columns. We then create a `univariate_analyzer` custom function to generate univariate analysis charts on our data. The parameters for the function are the `data` and the `subset` parameters. The subset parameter represents the data type of columns to be analyzed. This can either be numerical, categorical, discrete, or all columns. We then create conditional statements to handle the various subset values. Within each condition, we create a `for` loop on all columns within the data type category and plot the relevant chart for that category. We initialize a `matplotlib` figure in the `for` loop to ensure multiple charts are created and not just one chart. We then append our chart output into a list. Lastly, we perform univariate analysis on our dataset.

In steps 8 and 9, we create a subset of our dataset in preparation for bivariate analysis. This is critical because bivariate analysis on all 30 columns will generate over 400 different charts. We also identify various data types within the data since this is a requirement for bivariate analysis, given various chart options.

In step 10, we create value pairs for the various data type combinations. These combinations include numerical-categorical, numerical-numerical, categorical-categorical, discrete-categorical, and discrete-numerical. For combinations that involve different data types, we use a list comprehension to loop through the two categories and provide the column pairs in a tuple. For combinations involving the same category, we use the `combinations` function in the `itertools` library to prevent duplications. For example, (`'Income'`, `'MntWines'`) is technically the same as (`'MntWines'`, `'Income'`).

In steps 11 and 12, we create a `bivariate_analyzer` custom function to generate bivariate analysis charts on our data. The function takes the data as the only parameter. We create conditional statements to handle the various data type combinations. Within each condition, we create a `for` loop on all column pairs within each data type combination and plot the relevant chart for that combination. We then append our chart output into a list. Lastly, we perform bivariate analysis on our dataset.

There's more...

We can create a reusable module from the functions we have created. We achieve this by creating all our functions in a new `.py` file and saving this file as `automated_EDA_analyzer.py`. Once this is saved, we can call this module by using the following command:

```
import automated_EDA_analyzer as eda
```

We can then call each custom function using the dot notation. This following is an example of this:

```
eda.bivariate_analyzer(data)
```

For the preceding code to work properly, we must ensure the `automated_EDA_analyzer.py` file is stored in the working directory. A sample `automated_EDA_analyzer` module has been provided along with the code for this recipe at `https://github.com/PacktPublishing/Exploratory-Data-Analysis-with-Python-Cookbook`.

Index

Packtpub.com

Subscribe to our online digital library for full access to over 7,000 books and videos, as well as industry leading tools to help you plan your personal development and advance your career. For more information, please visit our website.

Why subscribe?

- Spend less time learning and more time coding with practical eBooks and Videos from over 4,000 industry professionals

- Improve your learning with Skill Plans built especially for you

- Get a free eBook or video every month

- Fully searchable for easy access to vital information

- Copy and paste, print, and bookmark content

Did you know that Packt offers eBook versions of every book published, with PDF and ePub files available? You can upgrade to the eBook version at packtpub.com and as a print book customer, you are entitled to a discount on the eBook copy. Get in touch with us at customercare@packtpub.com for more details.

At www.packtpub.com, you can also read a collection of free technical articles, sign up for a range of free newsletters, and receive exclusive discounts and offers on Packt books and eBooks.

Other Books You May Enjoy

If you enjoyed this book, you may be interested in these other books by Packt:

Python Data Cleaning Cookbook

Michael Walker

ISBN: 978-1-80056-566-1

- Find out how to read and analyze data from a variety of sources
- Produce summaries of the attributes of data frames, columns, and rows
- Filter data and select columns of interest that satisfy given criteria
- Address messy data issues, including working with dates and missing values
- Improve your productivity in Python pandas by using method chaining
- Use visualizations to gain additional insights and identify potential data issues
- Enhance your ability to learn what is going on in your data
- Build user-defined functions and classes to automate data cleaning

Hands-On Data Preprocessing in Python

Roy Jafari

ISBN: 978-1-80107-213-7

- Use Python to perform analytics functions on your data
- Understand the role of databases and how to effectively pull data from databases
- Perform data preprocessing steps defined by your analytics goals
- Recognize and resolve data integration challenges
- Identify the need for data reduction and execute it
- Detect opportunities to improve analytics with data transformation

Packt is searching for authors like you

If you're interested in becoming an author for Packt, please visit `authors.packtpub.com` and apply today. We have worked with thousands of developers and tech professionals, just like you, to help them share their insight with the global tech community. You can make a general application, apply for a specific hot topic that we are recruiting an author for, or submit your own idea.

Share Your Thoughts

Now you've finished *Exploratory Data Analysis with Python Cookbook*, we'd love to hear your thoughts! Scan the QR code below to go straight to the Amazon review page for this book and share your feedback or leave a review on the site that you purchased it from.

`https://packt.link/r/1-803-23110-6`

Your review is important to us and the tech community and will help us make sure we're delivering excellent quality content.

Download a free PDF copy of this book

Thanks for purchasing this book!

Do you like to read on the go but are unable to carry your print books everywhere? Is your eBook purchase not compatible with the device of your choice?

Don't worry, now with every Packt book you get a DRM-free PDF version of that book at no cost.

Read anywhere, any place, on any device. Search, copy, and paste code from your favorite technical books directly into your application.

The perks don't stop there, you can get exclusive access to discounts, newsletters, and great free content in your inbox daily

Follow these simple steps to get the benefits:

1. Scan the QR code or visit the link below

https://packt.link/free-ebook/9781803231105

2. Submit your proof of purchase

3. That's it! We'll send your free PDF and other benefits to your email directly

www.ingramcontent.com/pod-product-compliance
Lightning Source LLC
Chambersburg PA
CBHW062047050326
40690CB00016B/3007